CURED

NATHALIA HOLT, PHD, is an award-winning research scientist specializing in HIV biology. Her research has led to major developments in the HIV gene therapy field. She has trained at the Ragon Institute of Massachusetts General Hospital, MIT, and Harvard; the University of Southern California; and Tulane University. She lives with her husband and their daughter in Boston, Massachusetts.

Praise for *Cured*

"In this accessible and fascinating account, Holt, a research scientist trained at MIT and Harvard . . . brings to light the remarkable early breakthroughs in treating a once fatal condition."
—*Publishers Weekly*

"A fascinating discourse on how medical science is zeroing in on an HIV vaccine after several anomalous triumphs . . . An astute AIDS retrospective blended with contemporary updates on aggressive medical strategies."
—*Kirkus Reviews*

"In this exquisitely detailed telling of medicine's desperate fight to control the HIV epidemic, author Nathalia Holt reminds us that all the best medical stories are foremost stories of people—their determination, their courage, and their ability, in the best of circumstances, to rewrite history in a way that protects us all."
—Deborah Blum, author of *The Poisoner's Handbook*

"Chronicles this remarkable story with clarity, grace, and, most important of all, humanity."
—Carl Zimmer, author of *A Planet of Viruses*

CURED

The People Who
Defeated HIV

Nathalia Holt, PhD

A PLUME BOOK

PLUME
Published by the Penguin Group
Penguin Group (USA) LLC
375 Hudson Street
New York, New York 10014

USA | Canada | UK | Ireland | Australia | New Zealand | India | South Africa | China
penguin.com
A Penguin Random House Company

First published in the United States of America by Dutton,
a member of Penguin Group (USA) LLC, 2014
First Plume Printing 2015

REGISTERED TRADEMARK—MARCA REGISTRADA

ISBN 978-0-14-218184-3 (pbk.)
CIP data is available.

Printed in the United States of America
10   9   8   7   6   5   4   3   2   1

For the Berlin patients,
TIMOTHY RAY BROWN and CHRISTIAN HAHN,
and all those living with HIV

Medicine doesn't always work like in a textbook.

—Heiko Jessen

Contents

Part III: Treating the Berlin Patients

Part IV: The Cure

Cast of Characters

The Berlin Patients

Christian Hahn—The first Berlin patient, a German who received early therapy and an experimental cancer drug

Timothy Ray Brown—The second Berlin patient, an American who received a stem cell transplant of HIV-resistant cells in Berlin

The Scientists

Heiko Jessen—Christian's physician

Gero Hütter—Timothy's physician

Julianna Lisziewicz—Key collaborator with Heiko Jessen

Robert C. Gallo—Codiscoverer of HIV, brought Jessen and Lisziewicz together

Bruce Walker—Discovered how Christian's cure works

David Ho—Leading supporter of early therapy for HIV

Eckhard Thiel—Hütter's department chief who made Timothy's transplant possible

Carl June—Researcher translating Timothy's cure into a therapy for all

Paula Cannon—Researcher translating Timothy's cure into a therapy for all

David Margolis—Researcher translating Christian's cure into a therapy for all

Preface

The needle pierced through my double gloves and into the tender skin of my finger. It was a quick, painless prick. I sat motionless at the hood, grasping the enormity of the situation. The lab I sat in lay buried under hectic Sunset Boulevard in Los Angeles, the animal facility at Children's Hospital Los Angeles. Below that busy, crowded street was one of the quietest places I had ever experienced. A land where the air was super filtered, the doors heavy, the people unrecognizable under their gowns, masks, and hairnets. I had spent so many hours alone in that lab. Countless late nights were spent under the hood, the only sound the frantic, creepy squeaks of a thousand mice.

Now, in front of me, in a vented, pressurized hood, lay one of those guileless creatures: a small white mouse. Its breaths were heavy in sleep. Encircling its nose lay a tiny clear plastic mask, providing the mouse with isoflurane, a powerful anesthetic capable of keeping the animal motionless while I performed the risky procedure. But the problem was, the animal was not motionless, at least not entirely. Just as I leaned in to inject it with a highly concentrated, lab-adapted aggressive strain of HIV, the mouse jerked. In that moment, the unthinkable happened: I accidentally poked my finger with the needle.

It was the third year of my doctoral studies. I was working on an innovative gene therapy approach to treat HIV. The idea was to knock out a gene that HIV needs for entering cells. We could knock out this gene in stem cells and then infuse them in a patient. All the immune cells that matured from the progenitor cells could then be resistant to HIV. The hope was that this therapy would create a functional cure the likes of which had been seen in only one man: the Berlin patient. At the time, we didn't know who he was. All we knew was that we wanted to replicate his experience in others who had HIV. To test the therapy, we were injecting those modified stem cells into mice. But these weren't just any mice. They were genetically engineered to have no immune systems of their own. When the human stem cells are injected, they develop a working human immune system, or, at least, as human an immune system as you can have inside a mouse. The exciting thing was, we could inject these mice with HIV itself. We didn't have to use some other, related virus. We could use the real thing. In our study, we took this a step further. We didn't want to cure just any HIV. We wanted to cure the most aggressive, pathogenic strain we could find. If we could cure that virus, we could have a cure for all strains.

But now, the fact that we were using an aggressive strain of HIV seemed like a huge mistake to me. Not only could I be infected with HIV but I could be infected with a super strain of the virus able to cause AIDS at a rapid pace.

I was completely alone in the windowless room, the hum of

the vent hood, a device designed to protect me from pathogens, filling my ears. I sat for a moment and looked at the mouse. My first instinct was to pretend that none of this had happened. I didn't want to admit to anyone that I could be so foolish. Protocol dictated that I should immediately call for help and then remove my gloves so I could wash the wound for fifteen minutes, using a special soap designed to kill viruses and bacteria. But what about the mouse? I had no idea what to do. This was made further ridiculous by the fact that I had written the safety procedure. A single line haunted me: "All HIV infections will be performed with a minimum of two people attending." A safety mechanism designed just for this moment. Under the conditions I had outlined, I shouldn't have to worry about the mouse because someone else would be there to help. I had gone against my own procedure.

I couldn't simply leave the animal there. Turning to my left, I looked at the mouse's littermates. They were all under anesthesia, sleeping peacefully in the cage. If they were left under anesthesia too long, they would die. So would this mouse lying in front of me.

These weren't just lab animals to me. I had been there on the night each of them was born. I had held their tiny pink bodies in both of my hands as I injected millions of human stem cells into a vein that ran down their cheek and was no thicker than a human hair. I had watched them nervously as they grew up, knowing that some of them would die. Now, three months later, I was about to inject them with a virus responsible for the deaths of millions of humans. My relationship with these mice was not simple. Every blood draw, every procedure, I cared for them. I used anesthesia when other researchers didn't bother. I didn't want them to suffer even for a moment. If they suffered for reasons beyond my control, I put them down, even though each one was priceless to me, representing weeks of work.

On the other hand, HIV works quickly when injected. While only a small percentage of accidental needle sticks occurring in a hospital result in an infection, my case was different. In the majority of needle-stick cases, the person whose blood is drawn is on antiviral drugs and therefore has no detectable virus in the blood. My case was the opposite. The needle I held contained a highly concentrated form of the virus, designed to maximize the infectious dose given to each mouse. Some studies have shown that the chances of transmission are reduced if the exposed person receives antiviral therapy within one hour. The clock was ticking.

Almost as if someone else possessed me, I calmly carried out my experiment that day. I acted as if nothing had happened. I quickly lifted the skin of the animal's belly before me, injected it, and gave the mouse the prescribed dose of HIV. Exactly like the dose I had just, accidentally, stuck in my own hand. Relieved, I threw the needle in a bottle of bleach and stopped the anesthesia flowing to the mouse in front of me. I gently placed him on his back in his cage, careful to make sure that no bedding was covering his nose, potentially blocking his airway. I watched as his cage mates smelled and prodded him with their whiskers. I waited a minute, observing his respiration leap from the slow drug-induced pace to his normal frantic breaths. He woke with a jerk, flipping onto his paws. He would be fine, but would I?

Inexplicably, I then finished injecting each of the remaining mice with virus, repeating the same procedure that had just resulted in my accidental needle stick. I cleaned the hood and put away the anesthesia equipment and my tools. I took off my gown, mask, hairnet, and booties. Then, as soon as I grasped the door handle, my hand began to shake. Once outside the mouse room, I began to break down. I washed my hand for a timed fifteen minutes, furiously rubbing the iodine mixture into the tiny puncture just barely visible on my finger. I left the labyrinth of the basement lab and stepped out into the warm California sunshine. Traffic swarmed around me as I crossed the street to the office of my academic advisor, Paula Cannon.

"I've had a needle stick," I told her. There was laughter seeping in from the hall as students joked with one another. Paula's response was characteristic: She was calm and collected. She made a quick call to the clinic and off we went. Inside the main building of Children's Hospital Los Angeles, Paula distracted me with funny stories, complained of her husband, and bragged about her kids. Having lost my mom some years before, Paula had become a surrogate mother. She was an advisor, a mother, and a friend all rolled into one. I loved her. She stayed by my side as I got the antiviral drugs I needed.

That next month, I would take a standard drug regimen that millions of people take worldwide to keep the virus in check. The drugs weren't cheap. The hospital paid $1,000 for that one month of medication. Even worse were the side effects. I spent the whole month sick to my stomach, constantly vomiting. The medication made me tired. I didn't feel like myself at all. That whole month, while I complained about the effects of an accident I had created, I kept thinking about the many people who take these drugs every day. Not for a month but forever. Not everyone gets sick from the antiviral drugs, but many do. There are those out there who have trouble not just with the side effects but also finding a drug combination that will bring their virus under control. Even worse are the millions who are infected but have no access to the lifesaving drugs I complained about having to take. This book tells the story of how we have developed a cure but also what we still need to do to bring it to the 34 million people infected with HIV worldwide.

I was lucky. I didn't get HIV. We changed the safety practices in the lab and, with help, changed the way we injected mice with the virus to ensure that no one else from our group would get a needle stick. I like to tell this story as it demonstrates my weakness at a moment when I needed strength. I'm about to delve into the lives of two remarkable men, both of whom have been cured of HIV. When discussing their lives, I tell the high points and the low points. As I discuss their weaknesses it seems only fair that I disclose at least one of my own. More than weakness, though, this experience changed how I look at HIV. It altered what was once an abstract scientific concept, a medical problem that needs to be solved, and turned it into a human dilemma.

Two ordinary men changed the way we approached a cure for HIV. Their stories and those of other people reported on in this book were collected through hours of personal interviews with patients, friends, physicians, and researchers. In some cases, particularly those concerning events from more than a decade ago, memories differed between the parties involved. In some instances I report differing accounts; in others I present the account that best fits the associated facts and documents.

A few individuals, including the first Berlin patient, have asked to remain anonymous and I have respected these requests, altering names and identifying characteristics.

Before 2009, HIV researchers didn't use the word *cure*. Even today, there are scientists who cringe at the word. It's important to define what we mean by *cure*.

In science we talk about two kinds of cure: a sterilizing cure and a functional cure. A sterilizing cure is just what it sounds like: It "sterilizes" the body. This means that no trace of the virus can be found. A functional cure, on the other hand, doesn't get rid of all traces of the pathogen. But a cure of either type means that the person does not require medication or therapy. Both types of cure mean a person doesn't have to worry about the virus growing in his body or destroying her immune system. It is also highly unlikely that the person could infect anyone else.

In people with a functional cure, lying hidden in their bodies, detectable only by the most sensitive tests, will be traces of the virus. Persons receiving this treatment are cured, but they will always carry a small piece of the virus with them. For most, it doesn't matter if the cure they receive is sterilizing or functional; they simply want to be cured. Both Berlin patients received a functional cure for HIV, which is to say they both still carry the virus inside them and always will. This may sound like a strange cure, but it isn't. After a child recovers from the rash and fever of chicken pox, the virus responsible, varicella zoster, lies dormant in the nervous system for her lifetime.

Viruses are unique in the disease world for their ability to live within us without causing disease. Before the discovery of the first virus in 1892, plagues were seen as simple cause and effect between microorganism and disease—you get infected, you get sick. This is epitomized in Koch's postulates, a set of four criteria formulated in 1884 for identifying the relationship between disease and microbe. These guidelines define the cause of a disease in simple terms of abundance and pure infectivity. While the criteria work well for diseases such as anthrax and other bacteria, they were written before viruses were discovered and fail us when we consider the world of viruses we now know. A poliovirus is capable of infecting thousands but causes paralytic disease in only 1 percent—you get infected and you may *not* get sick. We are only just beginning to understand our shared evolution with viruses. Human genomes are riddled with ancient traces of viruses—viruses that, once multiplied inside us, became trapped in our DNA and then passed down the generations. In fact, about 8 percent of our genetic material can be traced to the skeletons of old retroviruses hidden in our chromosomes. The idea that we can harbor a deadly virus within us without the threat of disease is the basis for the functional cure for HIV. Yet the cure the Berlin patients received is only half the story. The other half is what we do with their cure, how it inspires the people and medicine around us.

When discussing scientific research, it is impossible to include all studies that could be considered relevant. I've included research that experts in the field believe is most pertinent and exciting. While most of this research is published, some studies are still in early stages, and therefore the results come from conferences and reports. It's important to note that these types of data are not as established as that published in academic journals.

This book is about two unique and controversial medical cases. In an effort to keep the conversation balanced, I've discussed the scientific issues that make the cases controversial. When appropriate, I've featured the opinions of those who performed the research or whose reputations carry considerable weight. Sources for conflicting opinions can be found in the Notes section.

This book also includes frank discussion of how new therapies are brought to market and the difficulties therein. How to maximize the limited funding for science is a key issue currently being debated by the research community. Investment in novel therapies remains inadequate. While we would like to think that the hard part of curing HIV is the cure itself, the real challenge lies in getting that cure to the millions who are living with the disease.

One of the great beauties of science is that each study, no matter how small, takes a step in advancing the field. Similarly, this book rests on the shoulders of the many books, research articles, and reports that have come before it. Each carefully examined case in the story of the cure for HIV represents a piece of the puzzle. I have attempted to put the pieces together.

PART I

A Doctor, Two Patients, and Some Tests

Some day, all this will have to be developed, carefully printed, fixed.
            —Christopher Isherwood, *Goodbye to Berlin*

A Doctor, Two Patients and Some Tests

CHAPTER 1

The Good Doctor in Denial

The streets were bursting. The crowd at the March on Washington for Gay, Lesbian, and Bi Equal Rights and Liberation became overwhelming. Dr. Heiko Jessen was finding it difficult to remain calm. Over a million people attended the rally. It was a sunny, mild day in April 1993. The cherry blossoms were at the end of their bloom, filling the mall with soft, pink and white blossoms. The flowers fell from the trees like a fragrant snow, filling the streets with beauty. Jessen needed a quiet spot to himself. He found a solitary bench far from the talks and demonstrations. As he sat, thousands of miles away from his home in Berlin, one circular train of thought occupied his mind: Andrew. While Andrew stood only a few hundred feet away in the crowd, emotionally he was distant. Their relationship was faltering. Andrew had cheated on him, yes, but Jessen forgave him because he loved him. Now Andrew was complaining of a cold.

For most people, a family member complaining of a cold is thoroughly normal. For physicians accustomed to easing the

fears of their family and friends, a cold is certainly nothing to
be worried about. But Jessen is not like most doctors. As Andrew complained of a sore throat, lethargy, a fever, and then a
rash, Jessen became increasingly concerned. His thinking was
guided by his experience in his small clinic in Berlin. Every day,
he saw the same constellation of symptoms. He spoke to young
men who were battling what appeared to be the flu. Yet, in the
back of their minds was one event. A night spent with a partner
they had just met, a party they couldn't quite remember, a
struggle to put on a condom. Many patients were highly specific, detailing their exposure, able to remember the day and
hour they became infected. This is because an influenza virus
didn't cause their illness. It was often another, very different
virus.

In medicine, a prodrome is a set of symptoms that heralds
the beginning of a disease. These first symptoms are distinct
from the disease itself, with traits shared among similar pathogens. Viruses, for example, share a common set of prodromal
symptoms. Before we run a fever, get the chills, and become
hopelessly nauseated, we usually get an achy, tired feeling. This
feeling serves as a warning to our bodies, a signal that we're
about to get sick.

Some viruses, like shingles and other herpes, share similar
beginnings as the virus begins to invade. The virus first goes
through an incubation phase. Like an egg sitting in an incubator,
the virus hides in our bodies, waiting until it's ready to make its
presence known. It is rapidly expanding during this time, madly
replicating. Incubation can last anywhere from minutes to

decades, depending on the disease and the individual nature of the person infected. This stage provides the virus's chance to build itself up; it's almost as if the virus is training for the fight of its life. By the time it's ready to move to the next stage, revealing the first signs of disease, our immune system is already losing the fight.

HIV, like many other viruses, spends the short incubation period wisely. The virus makes millions of copies of itself, all before the body can properly identify it and mount a tailored attack. By the time acute infection begins, tens of millions of viruses have invaded, not only attacking our blood cells but entrenching itself in our tissues. The virus wipes out the immune system in the gut. It forms a long-lived reservoir in multiple organs, including lymph nodes and bone marrow. The virus hides out in "resting" immune cells, so called because they are no longer dividing. The virus integrates itself into the DNA of the cell and then goes dormant. When the cell wakes up again, years, even decades, in the future, the virus will wake up with it, insidiously using the cell to make more of itself.

These resting T cells are like rare gems in a mine of rocks. Despite their scarcity, HIV is able to discover them. In its isolated hiding place, HIV can remain undetected for decades, beyond the touch of antiviral drugs. The latent virus remains present but is not obvious, at least to the immune system or to our drugs. This is why we can't fully remove the virus with the therapies we have today. No matter how good our drugs are at tackling the virus, they cannot reach the hidden reservoir of HIV in the resting immune cells. Bob Siliciano, a researcher at the Johns Hopkins

University School of Medicine, describes the challenge simply: "You are stuck with the virus unless you get every last latently infected cell." Even if an HIV-positive person has taken antiviral drugs for decades, even if they've eliminated all detectable virus in the blood, once they stop taking the antiviral therapy, the virus comes roaring back, returning to the same high levels it enjoyed before any drugs were taken.

In less than one year, the virus becomes a part of our cells and ourselves. By the time we begin to feel the first mild symptoms of the disease, the virus has enacted wide-scale irreversible damage on our bodies. Yet we remain unconcerned, naively believing that all we're suffering from is the flu.

This is why Jessen was distressed to hear about Andrew's cold. Combined with Andrew's infidelity, the picture was worrisome. Jessen turned the facts over in his mind, doubting that he was right to be worried—was he just being overanxious about the man he loved? *This is the problem with treating your loved ones*, he thought. *You can't trust your own judgment.* While it's generally accepted that doctors shouldn't treat their family members, they frequently do. In the United States, more than 80 percent of physicians have prescribed medications for a family member. While Jessen knew he was breaking the boundaries of the patient-doctor relationship, he couldn't help himself. He knew it would alarm Andrew, but he had to speak openly with him. On the plane ride back to Berlin, he laid bare his fears. Andrew, nervous, agreed to an HIV test.

Jessen performed the test himself in his clinic in the gay neighborhood of Schöneberg, in the former West Berlin. The clinic took up the second floor of an ornate building built at the turn of the twentieth century in the opulent Beaux Arts style. The floor was divided into space for a clinic and an apartment where Jessen lived. Returning to medicine after German reunification in the early 1990s had not been easy. There were limited opportunities for doctors to establish their own practices. In a country with universal health care, the German government tightly regulates medical providers, including the opening of private clinics. Germany has since experienced a shortage of doctors, but in the early 1990s, there was a surplus, which made new clinics near impossible. Jessen managed to squeeze in his request just before the government temporarily shut down all new private clinic applications. Today, new clinics rarely open; instead they are handed off from one physician to the next.

For his practice, Jessen created his own medical training, outside the constraints of academia. He devised a combination of specialties that would cater to the specific health needs of gay men: primary care, infectious disease, and sports medicine. He was particularly interested in helping vulnerable gay teenagers with no place to turn, who could come to his practice for treatment, counseling, and understanding. Jessen had completed specialized training in infectious disease for obvious reasons. He included sports medicine because he knew gay men spend time at the gym and subsequently had sports injuries. He found like-minded doctors to join his new practice, including a counselor trained to attend to the psychological needs of his patients.

Renovating the old apartment into a new, fresh clinic with the modern aesthetic Jessen wanted proved to be a challenge.

Ever the family doctor, Jessen made house calls in the neighbor-hood during the protracted renovation. Jessen's parents, who lived on the family farm in Northern Germany, came down to help. It took three months of scraping, plastering, and painting to get the walls of the clinic ready. Jessen's supportive family was always close by, one way or another. A few years later, Arne, Heiko's brother, would come to work for him as a physician in the clinic.

Heiko Jessen had grown up on his family's farm, working with cattle after school and summers. Because Jessen was the eldest son, his grandfather was adamant that he one day take over the farm himself. Their small village had even celebrated when Jessen was born, for his birth was seen as extreme good luck, a son to carry on the family tradition. Yet his father had other ideas. Since he had been pressured into becoming a farmer like his dad, he wanted his son Heiko to find his own path.

After Jessen finished his medical school studies in Berlin, he moved to San Francisco for his medical fellowship and to see what the United States had to offer. In much of the world, HIV was ending lives in rapidly growing numbers. This was espe-cially evident in San Francisco in the late 1980s. Young male patients, desperately ill, overwhelmed hospitals, where no ef-fective treatment could be given. The scenes were hopeless.

For Jessen, a young, gay physician, it was simply too much. It was the first time he had been exposed to the enormous im-pact of HIV in the gay community. In San Francisco, Jessen says, "gay life meant HIV." He felt himself withdraw from med-icine. The sight of so many young men destroyed by disease made him question why he wanted to practice medicine in the first place. One thing he was sure of: He had no future in treating HIV patients. He couldn't handle it. He returned to the German countryside, feeling confused about his future. Should he take an easy path? He considered becoming a country doctor. The simplicity of being settled near his family's farm was tempting.

This all changed in 1989, the moment he heard that the Berlin Wall had fallen. He immediately packed his bags. Part of this rush to return to Berlin was to take part in the enormous cultural experience and celebration of his city and country. For Jessen and others flocking to the city, Berlin had become "a big party; in the East everything fell apart, there were no rules, no rent. . . . It was the perfect escape from being a doctor." When he was once again in Berlin, Jessen relaxed into the party scene. For six months, he stayed away from medicine, filling his days with friends and parties. Beneath a sea of celebration, he tried to numb his mind to the horrible cases he had seen in San Francisco. It seemed to be working. The young, ambitious, brilliant physician was able to pursue a life outside a hospital. Here was an urban gay culture that was far from the culture of fear and despair he'd found in San Francisco.

Eventually, he rented a small apartment in the Schöneberg neighborhood in the former West Berlin. The neighborhood was quiet compared to the raucous parties and the squatter apartments of what used to be East Berlin. The neighborhood was leafy green, full of tree-lined streets. Small neighborhood gardens sat nestled between old, ornate apartment buildings. The neighborhood still bore scars from World War II: delicately sculpted, Baroque-style buildings stood next to newly constructed monstrosities sporting ugly, flat facades, the result of speedy patch-up jobs following the end of the war.

One night, at yet another boisterous party, he met a young American. The party was held, as so many were, in a squatter's apartment, which still held remnants of former occupants, former lives behind the Iron Curtain. As Jessen worked his way through the crowd, Andrew suddenly stood out. The American looked like a teenager in high school. His youthful complexion and bright eyes suggested a happy-go-lucky personality. The child of West Coast liberal parents, Andrew was charismatic, spontaneous, and adventurous. He embodied the complete opposite of Jessen's measured personality. That night Jessen, as he describes it, met the love of his life. This one man would rouse Jessen to pursue an HIV therapy without precedent.

There's no sign outside the door of Jessen's clinic. Instead, a humble sign in the window lets you know there's a medical clinic within. You enter the building through a dark, dirty vestibule. In front of you an aged staircase, dusty and winding, with no natural light, leads four flights up to the clinic entrance. The stairs serve as a fearful anteroom for patients expecting bad news. These were the steps Andrew climbed as he met Jessen at his clinic and home. They had returned from DC a few weeks earlier. The clinic was open seven days a week and didn't close for holidays. Andrew always knew where to find Jessen. The wall that separated clinic from home was more a thin membrane; it was impossible to keep Jessen from his work.

Jessen broke the news to Andrew. He had given so many HIV diagnoses before, to so many young men like Andrew. He was gentle, as usual, but this was different. He was diagnosing his own boyfriend, his partner, the man he loved and trusted. They held each other in Jessen's apartment, tears streaming down both their faces. It was 1993. Everyone with HIV died of AIDS. There was only one drug available to treat HIV—AZT (azidothymidine)—and it wasn't able to keep people alive.

Jessen's thoughts immediately raced to the researchers he knew and to an upcoming conference. He would do anything he could to keep Andrew alive. In the back of his mind, he considered his own risk. He'd had sex with a man who was HIV-positive. Intellectually, he knew he should get tested, but he pushed this thought down. He rationalized his reluctance by saying to himself that Andrew needed him right now. He would think about testing himself after he found a therapy for Andrew. Despite the fact that he was a physician, that he knew how deadly the virus was, he remained stuck in his denial.

Andrew felt lucky to have Heiko Jessen in his life. His friends, however, were not so trusting. They believed that Jessen was making up the diagnosis. They thought this doctor was manipulating the result in an effort to exert control over Andrew. Even after another physician confirmed the diagnosis, the friends remained paranoid. They tried to convince Andrew that it was a conspiracy, even going so far as to say that activists from ACT UP, the influential AIDS advocacy group, had infected him. Despite these influences, Andrew believed and trusted Heiko. He was HIV-positive. Since Jessen was going to push the boundaries of existing HIV medicine, that trust was about to get the ultimate test.

But Andrew was never to become one of the Berlin patients who are famous in research circles. He left both Jessen and

Germany. Andrew's gift was the passion he gave Jessen to pursue a novel and risky strategy against AIDS. Jessen's experience with Andrew fueled his commitment to be a new kind of family doctor, one who had the courage, the audacity, the temerity to pursue a cure for HIV. This passion would lead him to treat two men who would change medical history and would, in the process, each receive a title befitting a mystery novel: the Berlin patient.

CHAPTER 2

A Visit with the Family Doctor

In 1996, Christian Hahn walked through a crowded Berlin neighborhood on a beautiful sunny day in early summer, the kind of day when it is hard to imagine anything bad could happen. The streets were filled with people sitting outside cafés. Music streamed from bars and clubs despite the early afternoon hour. Christian made his way to the small, unassuming clinic popular among gay men. The month before, on May 10, he had done something foolish. He had been at a party. He'd had unsafe sex. Then, in the week that followed, he became ill. The infection made him sick and very tired. He had a sore throat and swollen lymph nodes. It reminded him of chicken pox. He believed he was infected with a virus, but not HIV. As the sickness raged, he decided to make an appointment with his family doctor: Heiko Jessen.

He met Jessen at his bustling clinic and told him his fears, detailing with precision the exact dates of his unprotected sex and illness. Heiko Jessen wasn't just anyone to Christian. He was his doctor, but like many of Jessen's patients, he also considered him to be a friend. Jessen was close in age to Christian, both in their late twenties. Jessen had opened his clinic five years earlier. Having lived through his own HIV scare, he was better able to empathize with his patients. He still thought about Andrew, the young man who had broken his heart and who was healthy, living with HIV and a new boyfriend in Spain. Four months after Andrew's positive test, Jessen took his first HIV test. It is hard to understand why a physician who knows the importance of treating early would wait to test himself, but love can be illogical. The test was negative.

Jessen's waiting room seemed to be constantly bursting with patients, and he found himself ending each day completely exhausted. Yet he thrived on the intangible rewards that came from directly helping his patients. He told his friends and colleagues that he wasn't focused on research; he wanted to work with people.

As soon as he heard Christian talking about his flu-like symptoms, his mind jumped to HIV. Among the young, gay clientele of his practice, Jessen saw a lot of the virus. It was common for him to be suspecting HIV in a patient.

Sitting at the desk in the exam room, Jessen spoke in a clear, calm voice. "We'll test you," he said, making eye contact with Christian. "We'll draw some blood. Tomorrow you'll come back and we'll talk. In a week we'll have the results." Jessen didn't rush him. He answered Christian's questions the best he could. It was obvious that Christian wasn't worried; he didn't believe he was infected. But as Jessen saw it, Christian's exposure matched up perfectly with his flu-like symptoms. It was exactly what acute HIV infection looked like. Jessen started his pretest counseling, which he gives all patients who could be infected with HIV, to prepare them for the results, whatever they might be. This counseling is a mix of empathy, science, and prevention whipped together to emotionally prepare a patient for the diagnosis. It helps patients recognize risky behavior and to understand how the HIV test works and what will happen if the test is positive. Jessen has become expert in ways to speak to patients about their HIV diagnoses. He even teaches a course, Breaking Bad News, to medical students at Humboldt University of Berlin.

There are two tests used in the clinic to detect HIV: The first detects our body's response to the virus; the second detects the virus directly. The first kind is an antibody test. Our body makes antibodies as a way to snare invaders. The problem with this kind of test for HIV is that it takes time for our body to make antibodies against HIV, on average 25 days. A month is a long time to wait for an HIV diagnosis.

While most people in the 1990s received an antibody test for HIV, Jessen decided to order a PCR-based (polymerase chain reaction) test for Christian. Jessen knew that, if his hunch was right and Christian had been infected, it would be too early for his body to begin making antibodies against the virus. Instead, he would have to test for the virus directly. The only way to measure the virus itself, instead of the body's response to it, was a PCR test. Performed in the lab, PCR-based tests identify specific parts of HIV, the genes found in every virus. The process causes the virus to multiply so that it can be detected in very small numbers. Ordering this test was unusual at the time; most physicians simply waited a month and then used an antibody test. Jessen, however, was influenced by the emerging research from David Ho's laboratory. Ho, a researcher in New York City, had a theory that initiating treatment early in HIV infection was key. Therefore, HIV diagnoses had to be made early. Jessen asked Christian to return for the results of the new lab test.

Christian had close friends and a warm, loving family. But he told no one he was being tested for HIV. Since he didn't believe he was HIV-positive, why mention it to anyone else?

Christian had grown up in a rural area in Southern Germany. He remembers his childhood as a happy one. He was interested in pursuing linguistics and began studying the topic at a school in his hometown before deciding to move to the capital city in 1995. Christian loved Berlin. The city was so different from his small hometown of ten thousand people. Although shy, he began to make lots of friends at his university. His social life was exhilarating, as it so often is for newcomers to city life. Then, at a party, he did something risky—had sex with a man he barely knew. Now here he was, only a year after moving to the big city, getting his first HIV test.

Despite the media attention HIV was receiving in Berlin, Christian felt insulated from the epidemic. He knew no one who was HIV-positive. "There was a man on the cover of *Time* magazine," Christian recalls, referring to David Ho, the person who would influence his own cure but whom he can barely remember. While the German newspaper *Der Spiegel* was reporting the increasing cases of HIV worldwide in 1995, it was hard for Christian to connect the dire statistics he read about with the reality of his group of healthy young friends. As he sat in the clinic waiting room a week after his first visit, and heard his name called, he felt relaxed and confident when he stood up to see his doctor.

Jessen told the nurse to prepare tea. It was a sign to the nurse that he was about to deliver an HIV diagnosis. He liked to have everything ready to comfort the patient as soon as he was done talking with him. Jessen entered the exam room and shook hands with Christian. He sat in a chair on the other side of the desk just a few feet away from the exam table. The room was all white. White walls, a white exam bed, thin, gauzy white curtains that ballooned out from the large open windows. The style of the furniture was modern, chrome and dark wood.

Jessen started by giving what he calls a "warning shot" before he broke the bad news. "I don't have good news for you," Jessen said to Christian, looking him straight in the eye. And he paused a few beats.

This would hardly be the only HIV diagnosis he would give that day. His waiting room was crowded with infected young men. But Jessen was especially gentle with Christian, even when he said simply, "Your HIV test came back positive." He wanted to impress upon him the seriousness of the virus, but he also wanted to assure him that there was treatment available and that it was critical to start the drugs immediately. With the bad news, Jessen embraced Christian, rubbing his back to soothe him.

While breaking the news to Christian, Jessen assessed his patient's personality. Jessen was already unusual among family doctors in having begun to combine research with his practice. As he sat with Christian, he wondered if he could benefit from an experimental new drug, a drug he was giving sparingly to a small group of patients early in their infection. Jessen had been the first to give the drug to an HIV patient. The first patient was his boyfriend, Andrew. Now Jessen wondered if the drug had the ability to help people worldwide and if Christian, a responsible young man, could be part of this trial.

Christian felt eerily calm as he heard that he'd tested positive. He couldn't wrap his brain around the reality of the situation. He was unable to comprehend what he was told and unable to process the fact that his life was about to radically change. It was as if Jessen were speaking in a foreign language, and Christian struggled to understand the meaning. They talked. They paused. They talked again.

Jessen was patient. He had become accustomed to volatility in patients' reactions. But he was already thinking Christian could be counted on to take his medicine reliably.

"We're testing a new drug that might be able to eliminate HIV," Jessen began.

Christian only nodded, asking no questions. He would take whatever Jessen told him to take. Jessen's idea was to initiate therapy early with an aggressive drug regimen, before the virus could take hold in the body. For this trial, Jessen selected only patients who were very early in infection and whom he trusted to be reliable.

He reviewed with Christian the drugs he would need to take, and when. The whole idea of combining drugs was new in 1996, and Christian would be taking three: didanosine, indinavir, and hydroxyurea.

When scientists working at Roche, then one of the world's largest biotechnology companies, solved the crystal structure of

HIV's protease enzyme in 1989, it cleared the way for a new strategy to treat HIV. Multiple pharmaceutical companies used the structure of the viral enzyme to design a new, effective class of drugs. One of those was indinavir, available from Merck and approved three months after Roche's inhibitor, saquinavir. This was the drug Christian would take.

Unlike the brand-new protease inhibitors, hydroxyurea was an experimental drug that only a handful of people worldwide were testing, and Jessen thought it might be capable of purging the virus from the body. The problem with all these drugs was that they lasted only a short time in the body. This meant that Christian would have to take didanosine twice a day, indinavir three times a day, and hydroxyurea three times a day. The drugs had to be taken at precisely the right time in order to keep the virus in check.

Jessen explained that they would need to take another blood sample for further confirmation and that Christian would need to come back the next day. He wrote Christian a certificate that gave him two weeks off from his school internship. He then sent Christian to the tearoom, where he would talk to a nurse for two hours and drink tea. And that was that. Years later, Jessen said, "I had no indication of how special he would be."

Christian, for his part, did not feel special at the time. He folded the instructions in his hands and shuffled out of the office. He wasn't sure how he was supposed to feel, but he was certain he was not having the right reaction. He wished he could cry or moan or be somehow demonstrative. Instead he felt numb and alone. That night, he would write out the complicated schedule of drugs he was about to take. Jessen had convinced him that these drugs were key. All he had to do was adhere to his schedule. And never miss a dose.

What neither of them knew is that, a year later, Christian would become an anomaly. Researchers would start writing about him as "the Berlin patient" and his story would change the HIV field forever. At the same time Jessen was putting Christian on an experimental new therapy, another one of his patients was receiving a standard treatment regimen. That man, diagnosed later in infection, was three years older than Christian and also living in Berlin. Although they never met, their early experiences in being diagnosed with HIV were eerily similar. They spent time with some of the same people, went to the same doctor, hung out in some of the same clubs and restaurants. That man, Timothy Brown, would also become known as the Berlin patient, although he was the second such one. These men, despite the very different therapies they received, would share in a unique experience: They were both cured of HIV.

CHAPTER 3

Death Sentence?

Sweat dripping down his face and back, Timothy Brown leaned against the brick wall. He was excited and out of breath, his heart pounding frantically in his chest. The deep beat of techno music seemed to pulse through the wall and pull him back into the packed club. In the 1990s, Tresor was a legendary underground techno club. It was the place to be in Berlin, with lines often stretching down the block. Located in the center of the former East Berlin, its name, *tresor*, roughly translates as "the vault." This one, unlike the crypts once concealed in Gothic cathedrals, was located in the defunct bank vault of a gone-out-of-business department store.

Timothy loved Berlin; he loved the nightlife, his friends, his boyfriend. Life couldn't seem to get better for him than it was in Berlin in 1995. The place was experiencing a rebirth as people from all over the world flocked to the reunified city. In Christopher Isherwood's memoir, filled with his experiences in Berlin in the late 1920s and early 1930s, he writes, "Berlin meant boys." Berlin in the 1990s was reminiscent of that earlier time of sexual liberation, before World War II. Timothy was experiencing that kind of freedom.

People who knew Timothy found him terribly charming. He reveled in being with his friends, flirting with everyone, and filling up small spaces with his staccato laughter. He was a student in Berlin, still unsure what he wanted to study. To pay the bills, he worked at Café Einstein, located right at the foot of Checkpoint Charlie. This famous Berlin Wall crossing point, where travel was once restricted between West and East Germany, was now a bustling tourist destination. The café was constantly packed with travelers.

Standing outside Tresor that summer night, Timothy thought about Marcus, an old boyfriend he'd dated for six months nearly two years earlier. Marcus had been constantly jealous, always imagining that Timothy was off chasing other men. They were traveling in Greece when Marcus abruptly broke up with him on the island of Mykonos. Timothy was miserable when Marcus left him. Now, waiting on Timothy's answering machine was a message from the man he still thought about. Marcus wanted to see him.

In the face-to-face meeting, though, Timothy's daydreams were shattered. Marcus wasted no time in getting to the point. "Hey, I got tested for HIV and I tested positive," he said. "You need to get tested, too." Timothy didn't take the news lightly. This was the mid-1990s and he knew HIV was the leading cause of death in Americans age twenty-five to forty-four. The previous March, Timothy had said good-bye to a dear friend who had been diagnosed only a year earlier. Everyone who was diagnosed with the disease died, and Timothy had lost so many friends.

HIV was a death sentence. There was no good treatment, and no cure.

Timothy knew that just because Marcus was positive for HIV didn't mean he himself would be. In fact, Timothy was sure he wasn't HIV-positive. He had been pretty careful. However, one night stood out in his mind. A night before he'd ever met Marcus. Usually, when he had sex, he asked his partner to not ejaculate inside him. It wasn't a perfect solution, it couldn't protect him from the diseases he knew were out there, but it was better than nothing. One man he had been with, Jeremy, had blatantly ignored that request. It was to this man that Timothy's thoughts began to turn. He had seen him only once since that night. It was a casual meeting. As he spoke pleasantries, only one thought dominated his brain: *You're the guy that ejaculated inside me.* Was it possible that Timothy had infected Marcus, that he himself was HIV-positive?

A few weeks after Marcus announced he was HIV-positive and that Timothy needed to get tested, Timothy sat in a small, clean clinic located in the Institute of Tropical Medicine of Charité hospital. It was 1995 and the first time Timothy was seeing a doctor since he'd moved to Berlin. A decade later, after he was diagnosed with cancer, he would come to know the hospital well, its rooms and walls becoming like a second home to him. But this was his first time here, and he had trouble finding his way in the labyrinth of halls. As he sat in the waiting room of the tropical medicine clinic deep within the belly of the hospital, he realized he was feeling the same anguish he'd felt in the clinic back in Seattle when he got his first HIV test years before. The waiting was terrible. It had taken weeks to get the results.

While the kind of test Heiko Jessen gave Christian Hahn is often used today to detect early HIV, in 1995 these tests were new and rarely used. Instead, the common test was the ELISA, or enzyme-linked immunosorbent assay. It determines whether or not our immune system is responding to the virus. That is, the ELISA detects the antibodies our body makes against the virus. Any pathogen that invades our body makes itself known to the immune system by showing little pieces of itself on the surface of the cell it has invaded. These pieces of a pathogen, called antigens, invoke a response from the immune system. Once the body detects the antigens, distinct for each virus and bacteria, it mounts an immune system attack.

The immune system attack comes in two waves. The first wave is from the innate immune system, which is composed of a farrago of deterrents. It includes cannibal-like cells that eat other infected cells, and inflammation that creates a physical barricade between the infection and the rest of the body. The innate immune system can be soldiered up quickly in response to a pathogen because it uses the tools already on the shelf.

However, the second wave of the immune system attack, from the acquired immune system, takes more time. The acquired system develops new weapons designed for the invading pathogen. To mount its attack, the acquired system uses the blood's infection-fighting white blood cells, specifically lymphocytes, comprised of T cells and B cells. T cells are named for the location where they become differentiated from stem cells, in the thymus, and B cells for their differentiation in bone marrow. For HIV, this "custom-made" immune response takes time, from weeks to months. The average time is about 25 days.

If you come upon a nail and then invent a hammer so you can use the nail, you probably wouldn't throw away the hammer after you use the nail. After all, you might find another nail. Just so, once the B cells make those antibodies for HIV—or any virus—the infected person's immune system will always remember the virus and continue to make those same antibodies, just in case, for the rest of the person's life.

To perform the ELISA test, we take a small amount of purified blood, which is called plasma and is a translucent yellow. We dilute it several hundredfold, and put the diluted plasma into the well of a so-called 96-well plate. This plate, made of clear plastic, is molded with 96 tiny dimples, or wells for holding liquid. The amount is tiny, the size of a few raindrops. Inside each well is an antigen, a piece of the virus just large enough to attract the attention of the still functional immune system contained in the well. If the immune cells immediately recognize the virus and attack it, the test is positive, indicating that the person from whom the blood was taken is infected with HIV.

But how do we know if the immune cells are attacking? We don't need lab technicians to peer through a microscope at what is going on. In the case of a positive result, when the person's antibodies bind to the foreign intruder, the fishing line is reeled in. The antibody is captured on the hook and a brilliant purple dye on the other end of the fishing pole is released. The well turns purple. The darker the color, the stronger the immune system response. If a person's diluted blood plasma makes no antibodies to HIV, they have never seen the virus before and no purple dye is released. Go fish.

The ELISA is labor-intensive, requiring skilled technicians

to prepare the materials, load them onto the microplate, carefully wash the plate, and read the results. It is a very sensitive test: It accurately diagnoses HIV in 99.9 percent of those infected. But there are two big weaknesses. As we've seen, it takes our bodies time to make antibodies against HIV, so a person can be infected for months and still be negative on an ELISA. For this reason, ELISAs aren't typically given until at least six weeks after an exposure to HIV. The ELISA is very accurate if you've made antibodies to the virus, but not accurate at all if it's too early after infection. The second weakness is the amount of time the test itself takes to run—about two weeks. These are two weeks full of torment. In 1995, the year Timothy tested positive, it was estimated that approximately one-third of patients who tested positive for HIV did not return for their results.

Today you can buy a rapid HIV test, called OraQuick, at a local pharmacy. This test is like holding an ELISA in the palm of your hand, but even better, since it requires no blood from you. Using a swab, you collect mucosal transudate from your mouth. Unlike saliva, which comes from a gland in the mouth, mucosal transudate resides in the cheeks. This clear fluid mixes with saliva in the mouth, the two indistinguishable. The transudate, however, as it comes from the cheeks, is enriched with antibodies. While there isn't any measurable HIV in the mouth, the antibodies that every HIV-positive person makes are released into their tissues and blood. These are what the test is swabbing for. The swab is placed in a vial containing replica pieces of HIV. These replicas look like HIV but they can't infect anyone. If they are present, antibodies attack them. This causes a color-changing enzyme to react. In about 20 minutes, a line will appear in a box on the device (much like on home pregnancy tests that so many women have inspected with emotions ranging from hope to fear). On the OraQuick test, if a second line appears, you are HIV-positive. The home test is very accurate although not quite as good as the laboratory version. Today, all this can be done in your living room, but in 1995, the long wait ended with a potentially life-changing visit to the doctor. Jessen worries about the downside of home HIV tests, though, saying, "It's not news that anyone should have to deal with alone." His opinion is shared by other physicians, particularly family doctors, who believe that counseling is an important component of an HIV diagnosis. It's a difficult balance between making HIV testing convenient and making sure that patients receive the support they need. Apart from the emotionally traumatic aspects of an HIV-positive diagnosis, new testing technologies came at just the right time, making possible a critical step forward on the road to the cure of HIV.

The tropical medicine institute at Charité hospital was time-worn, with faded paint and old furniture. Patients describe the clinic as dark, with only a soft gray light filtering in from small upper windows. A poster on the wall read AIDS GEHT ALLE AN PROBLEM ("AIDS Is Everyone's Problem") and presented black-and-white photographs of men and women, their heads bowed as if in prayer. Timothy was called into a small room. He shook hands with a doctor who held his test results in his hand. Timothy's lips trembled.

On hearing a diagnosis of HIV, some patients seem to know or suspect, some can give the exact time and place they were infected, and others are shocked and never saw it coming. The immediate impact of the diagnosis, whether a person falls apart or is brave for the medical staff, is like a snowflake, as unique as the person.

Timothy wanted to tell everyone he knew. He told his boss at the Café Einstein. He told his coworkers. He told his friends. He says, "I did not want to be quiet." Over and over again, he spoke aloud those terrifying words he had heard from the doctor. He was HIV-positive. Among all those he told in the first few weeks, two people were notably absent. The first was his mother. His mom was ill with breast cancer. He felt he couldn't add this burden to her life. If he told his mother, he knew she would be scared for his life.

When Timothy's mother met his father, she was a teenager, captivated by an older man she barely knew. She was a Christian woman from a strict family, overcome by teenage hormones. She was shocked to learn that not only was she pregnant but the father of her unborn child was already married with children of his own. He left her. Timothy grew up without a father. Timothy's relationship with his family was fragile. It seemed the family could absorb great pain, each person buckling under the weight but still holding on.

The second person he couldn't tell was the man he believed infected him. Jeremy. Most physicians will tell their patients to alert all the partners who could possibly be infected and especially the person they believe infected them. This is a public health service, to keep those who are infected from unknowingly spreading the disease further. Timothy wasn't sure where Jeremy was; he couldn't even be sure he was the man who had infected him. He had seen him only once since that night some years before. Timothy could be fearless, but something about Jeremy made him recoil from finding him.

Of the many people he did tell, the first person was his boyfriend at the time. His boyfriend's reaction was extreme. He immediately burst into tears of anger. "You'll be dead in two years," he said, pounding his fists on his legs. "Your life is over."

PART II

The Disease, a Drug, and Its Industry

This was a time in which the United States boasted the world's most sophisticated medicine and the world's most extensive public health system, geared to eliminate such pestilence from our national life.
—Randy Shilts, *And the Band Played On*

CHAPTER 4

Viral Trojan Horse

AZT is a hot button for many people living with HIV. It's an ugly part of our history, a drug that revealed unfairness in our drug development industry, homophobia in our government, and a lack of empathy in those who should have been protecting the public. HIV researchers tend to have a different view of the drug. For them, it represents the first hope, the forefather of all HIV drugs to come. Today, it's one of the few drugs considered safe for an unborn child.

In 1984, Burroughs Wellcome Company, the company that came to make windfall profits from AZT, was the twentieth-largest pharmaceutical company in the United States. David W. Barry, vice president of research, specialized in viral diseases. It was a time when few companies researched viruses, which are notoriously difficult to target. Unlike bacteria, viruses form an intimate connection with an invaded cell's machinery. Similar to cancer, it is almost impossible to kill a virus without killing the cell as well. Barry was particularly interested in AIDS, the new disease that had been discovered only three years before.

Although it's hard to imagine a pharmaceutical company taking such risk, in 1982, Barry formed a small group to begin considering what drugs could treat the new disease, then briefly known as gay-related immune deficiency, or GRID, which had no known cause. It was a bold move. No other pharmaceutical company would remotely consider devoting resources to this disease. Barry found himself drawn to working on the disease despite, or perhaps because of, the enormous challenges it presented. Those closest to Barry found his interest in AIDS bordering on obsession. Early in 1984, French and American scientists had, at almost the same moment, established that AIDS patients were infected with a retrovirus.

HIV is the retrovirus they identified. It is a kind of viral Trojan horse. It is "retro" because it reproduces itself in a way that's backward from how most life-forms manage this wonderful trick. In the nucleus of each cell in our body, and we have an awful lot of them, the genetic recipe for building our entire body can be found. Every cell has the complete list of instructions. Those genes are packed in DNA, tightly coiled molecules of nucleic acids. To make use of all the instructions in our genes, certain enzymes unwind the DNA helix in the cell's nucleus. But to ensure that the precious DNA is not lost, another enzyme swoops in and makes a copy of the DNA needed. This is called transcription. This copy is made literally backward and has a few special alterations. (One of these alterations is to a base called uracil, a molecule whose origin may be extraterrestrial—yes, really.) The copy is called ribonucleic acid, or RNA. The RNA blueprint is then transported from the nucleus to a ribosome, which, once it gets its hands on the RNA, translates the backward code and makes proteins from the blueprint.

DNA → RNA → Protein

Simple as one, two, three. For decades, this definition of the process, made by Francis Crick, the codiscoverer of DNA with James Watson, was considered scientific dogma. That is how life works. So imagine how surprising it was when scientists discovered that a virus could run in the opposite direction. Retroviruses forced scientists to question what defines life. Here was an organism with a genetic code of its own but with no cell to house it. Can you be alive if you hold the blueprint to life but have to borrow the construction plant from another organism? Can there be life outside a cell? Taxonomists have spent decades debating this, perhaps pointlessly. As Carl Zimmer argues in *A Planet of Viruses*, "Trying to find a moment in time when such RNA-life abruptly became 'alive' just distracts us from the gradual transition to life as we know it."

This is what retroviruses do. They trick our cells into making the proteins they need, as directed from their RNA. Instead of starting with DNA, HIV holds all its genetic information as RNA. The virus exists as two simple strands of RNA packaged into a spiky protein shell containing all the enzymes it needs. The viral RNA travels into the nucleus of our cells and instead of performing transcription, it performs reverse transcription: HIV uses an enzyme called reverse transcriptase to transform a copy of its RNA into DNA (the same format as the DNA in our own human cells). By converting its RNA to DNA, it can insert its genetic material into our own. Because all the virus's genetic

material is now DNA, our immune system can't distinguish between viral genes and our own human genes. Once it has done
this, the virus has essentially tricked our cells into making the
proteins that HIV needs to make more virus.

RNA → DNA → RNA → Protein

After the virus has used reverse transcriptase to make a
DNA copy of its own RNA, it still needs to hide it in our DNA.
To do this, it uses another enzyme, integrase, to assimilate the
newly made DNA into human genetic material. Integrase cuts
into our DNA and then joins the cut strands of our chromosome with the newly made viral DNA. This step is irreversible;
once the process is complete, the virus will always have a place
in our chromosomes.

As directed by the virus, the invaded cell produces long, unwieldy chains that incorporate the viral enzymes: reverse transcriptase, integrase, and protease. Like a salad, these proteins
have to be chopped up and mixed together to make a virus. The
virus uses the enzyme protease to do this. Without protease, the
virus would have all its basic components but would not be infectious. After the protease finishes dicing proteins, the virus
undergoes its final assembly, bringing together single-stranded
RNA, viral enzymes, and core proteins to form a capsid, that is,
a protein shell containing everything it needs except the viral
envelope. The virus picks up this last piece of the viral puzzle as
it is leaving the human cell. The envelope—the proteins that
surround the virus—is part virus and part human. Because the
virus has this unique coating, it can go on and infect another
cell. This life cycle is shown in the illustration on page 37. A
mature virus particle, or virion, is born.

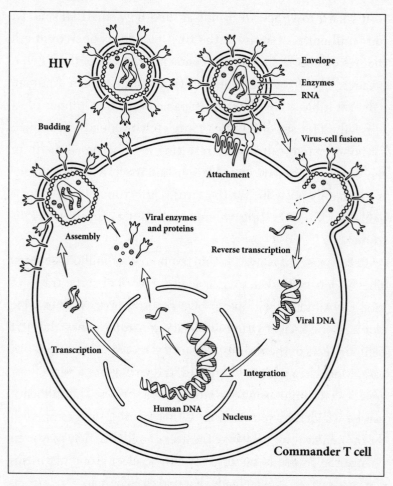

How HIV invades a cell. The virus first makes contact with a T cell, releasing its enzymes and RNA inside. The reverse transcription enzyme translates the viral RNA into DNA. The virus then makes its way to the nucleus, where the integrase enzyme hides the viral DNA within our human DNA. The viral DNA is transcribed back into RNA by the cell. Our cell then makes viral proteins as directed by HIV. The protease enzyme assembles these proteins into a virus. As the virus leaves the cell it picks up proteins from our cell membrane, giving it the key to unlock more T cells.

It's hard to wrap our minds around how small HIV is. At four-millionths of an inch, the tiny intruder is one-twentieth the size of a bacterium, one-seventeenth the size of the T cell it invades. The virus is a thousand times thinner than a human hair. Yet it has a big footprint, capable of making billions of itself daily. This sea of invaders completely overloads the human immune system, ultimately resulting in widespread death of the very cells the virus needs to sustain itself. Killing our cells is ultimately not wise for the virus. Unfortunately, by coincidentally killing the cells we need to protect ourselves, it kills us, too.

Retroviruses have lived within our bodies for millions of years. They've left behind an archaeological record of sorts, traces of viral DNA hidden within our own genome, impossible to wipe clean. As the ancient viruses invaded our chromosomes, they left behind pieces of themselves, a historical record of infectious disease. More than a historical record, the viruses are a part of our genetic code, influencing our progress as a species. This influence can be wielded by retroviruses. Other viruses, such as Spanish flu and yellow fever, may take the lives of millions; they may even change the course of history, but only a select group of viruses can conquer who we are and what makes us human.

We think of retroviruses as faceless monsters, destroyers of life. But not all retroviruses harm their host. What is the difference between the retroviruses that destroy and the ones that do no harm? The answer seems to be evolution. Two retroviruses that can live peacefully with their animal hosts are closely related to HIV: simian immunodeficiency virus (SIV) and feline immunodeficiency virus (FIV). In FIV infection of cougars, where the virus has a long shared history, the infection causes no disease. However, once transferred to domestic cats, where the virus has a much shorter evolutionary history, FIV can cause AIDS-like symptoms. A similar scenario exists in monkeys. Some species, such as African green monkeys, live in relative harmony with their SIV, harboring the virus with little consequence. These monkeys have likely lived with their SIV for millions of years, plenty of time for both the animal and the virus to find the right balance. Contrast this with the consequence of an SIV that was transferred to humans: HIV. HIV has had only a short time to get to know us, approximately one hundred years. We've known HIV even a shorter amount of time, about thirty years. If we could wait a million years, perhaps we could achieve a truce with HIV. Since viruses are ultimately driven by biology to make more of themselves, the best way to make more of themselves is to keep us, like the African green monkeys, alive and replicating. Ironically, the only way HIV can succeed as a virus is to allow our survival.

HIV is not one uniform virus. It swarms in our bodies but varies genetically from particle to particle. When the virus makes DNA from its RNA, the product is riddled with mistakes, and this gives it a distinct advantage in being able to adapt and mutate. The virus's poor ability to precisely copy itself makes it, as a swarm in the body, more resilient. That's why there is such a high level of drug resistance to HIV. Even though a drug may be able to efficiently attack one part of the virus, somewhere in that swarm probably exists a lone wolf able to evade the drug. That variant will begin to replicate until it can overcome the drug's affect. It is this unique property of HIV that makes it so difficult for us to develop effective drugs to fight it and the reason new antiviral drugs are constantly in development.

Almost every person infected with HIV, if they don't receive treatment, will progress to AIDS. HIV wears down the immune system, killing off our immune cells, particularly our CD4 T cells. Without these T cells, the commanders of the immune system, we are vulnerable to diseases we normally would be able to defeat. Everyone advances at a different pace; for some it may be decades, and for others only weeks. On average, for a person off therapy, it takes 10 years to go from HIV infection to AIDS. For this reason, AIDS is defined either by the loss of the commander cells—less than 200 of them in a microliter of blood (we normally carry between 500 and 1,000 per microliter)—or by the presence of an AIDS-defining illness. Defining illnesses, rarely seen in healthy people but common in persons with AIDS, include a bacterial pneumonia and a herpes-driven tumor that causes lesions across the body. The presence of one of these diseases indicates that the immune system has broken down and left the body defenseless. Beyond its clinical definition, AIDS carves a wide path in the body. Extreme fatigue and a wasting syndrome characterize the disease. Those with AIDS resemble cancer patients, with sunken cheeks and slim bodies. For those who evade death, the disease still carries unshakable stigma.

The discovery of the reverse transcriptase enzyme in 1970 flew in the face of everything scientists thought they knew about DNA. The discovery came from two independent research groups: geneticist Howard Temin with his postdoctoral fellow

Satoshi Mizutani at the University of Wisconsin–Madison, and biologist David Baltimore, a young investigator at the Massachusetts Institute of Technology. Fifteen years before the discovery of HIV, Baltimore was studying a lesser-known retrovirus, rous sarcoma virus, when he came across the unique enzyme. It was a watershed moment in virology. These men would share the Nobel Prize for their groundbreaking work only five years later. For Baltimore, it was just the beginning of his lifelong investigation of retroviruses. The discovery of reverse transcriptase was a pivotal moment for molecular biologists and, although they couldn't know it at the time, for HIV therapy. Because of this discovery, inhibitors for an enzyme essential to HIV were already in development thirteen years later when the new virus, HIV (initially named human t-lymphotropic virus type III, or HTLV-III), was found. But the road to effective pharmaceuticals would be a tortuous one.

CHAPTER 5

A Weapon from the War on Cancer

In 1936, there was some inkling that environmental factors could be contributing to cancer. Tobacco, radiation, hormones, and asbestos were seen as possible, although unconfirmed, influences. If World War II had not monopolized the need for scientists, some researchers believe these early findings would have been acted upon. Although impossible to know, perhaps our approach to cancer would have been more coordinated, more rational.

By the 1940s and 1950s, we were suffering under the weight of a huge number of cancer diagnoses, about 200,000 each year. We kept the disease shrouded in mystery as a matter of propriety. Very little money was spent on research. The media largely avoided the subject. In fact, the word *cancer* was prohibited from being said aloud on NBC (the National Broadcasting Company).

Much like the stigma generated by HIV, cancer was both incomprehensible and shameful. It was described as the "disease of civilization," blamed on modern life, even considered a punishment. Patients hid their disease, afraid to talk openly about their fate. Incurable diseases expose our vulnerability, inciting fear, provoking judgment, and rousing our worst instincts in society. At the same time, they inspire some individuals to make radical choices. This is what happened to Mary Lasker, a socialite in New York City.

Beautiful, charming, and rich, Mary was born to an extraordinary fate. Raised in an upper class family in Wisconsin in the 1920s, Mary became plagued with childhood ailments, from dysentery to recurring painful earaches. Illnesses that today can be treated with simple antibiotics caused Mary to be isolated and lethargic.

As a young girl, Mary remembered standing outside a shack at the edge of town with her mother. They were dropping off laundry. Mary's mother paused outside the door to warn her daughter before they entered the room. "Mrs. Belter has cancer and her breasts have been removed." Mary replied, "What, cut off?" Mary's mother nodded and they entered the room. Mrs. Belter was lying in a low bed, her bedclothes failing to cover the scars of her surgery prominent on her chest. Seven noisy, demanding children crowded around her reclining figure. The moment would become a part of Mary's identity. She would later recall, "I'll never forget my anger at hearing about this disease that caused such suffering and mutilation and my thinking that something should be done about this."

By the time Mary was in college, her father had become frail, hardly eating. Her parents suffered from hypertension, a disease that would take both their lives when Mary was in her thirties. This left her "deeply resentful" of physicians and medical research. She scorned the medical institution that had not been able to help her family. She would later look back at these early experiences as the inspiration for her work, saying, "I found out that it all went back to my violent reaction and hostility to illness for myself or for anybody else."

When Mary Woodward secretly married Albert Lasker, the advertising legend, on a June day in 1940, there was no reason to think she would become a powerful advocate for medical research. Or that her advocacy not only would impact cancer patients but would form the foundation for all the HIV drugs we have today.

In the year before she met Albert Lasker, a shift had occurred in Mary's life. A divorcée living in New York City, she was close friends with Margaret Sanger and began raising money for the Birth Control Federation of America, the predecessor of today's Planned Parenthood. Her eyes were opened to the inadequacies of the public health system.

Following her wedding to Lasker, she became concerned about the declining health of her longtime housekeeper. The woman, although obviously ill, would not specify what her disease was, so Mary resorted to confronting her housekeeper's doctor directly. He told her, in a time before privacy laws, that the woman was hiding uterine cancer. A short time later, Mary was horrified to learn that her ill housekeeper had been sent to an institution "called something like the home for the incurable."

Mary, now armed with strong allies, thanks to her marriage to one of the most powerful men in New York City, decided to take on cancer. Upon learning the minuscule size of the budget for the American Society for the Control of Cancer, a tiny organization with only a thousand members and no research agenda, she decided to reform it. She used a fresh tactic to bring the need for cancer research to the public: advertising. With her husband's help she was able to convince David Sarnoff, the CEO of NBC, to lift the ban on saying *cancer* on the network. So convincing were her arguments that he not only lifted the ban but also agreed to have Bob Hope deliver a message on-air about the dire need for cancer research funding. She convinced editors at *Reader's Digest* to publish articles about cancer, suggesting writers capable of stirring emotions on the subject, and on at least one occasion writing a piece herself for the magazine. All in all, she raised an incredible amount of money.

But, sadly, science failed her once again. In 1952, her husband died of colon cancer. Many women might have become embittered, angry at both the disease and those who claimed they could fight it. Not Mary. She returned to her fight against cancer renewed; her beloved husband's death only strengthened her advocacy and lobbying.

Renaming the small cancer organization the American Cancer Society (ACS), she successfully lobbied Congress for research funding for the National Cancer Institute (NCI), boosting cancer research funding from $1.75 million in 1947 to an astounding $110 million in 1961. By the early 1960s, half of this enormous budget was spent by the NCI on screening thousands of possible drugs and sifting through a sea of chemical compounds to identify a few drugs with promise. Although this newly defined "war against cancer" was progress, the approach left much to be desired. The science of throwing random compounds and drugs at mice with leukemia was not innovative or exciting.

Jerome Horwitz, who worked at the Michigan Cancer

Foundation, part of Wayne State University, in Detroit, was a young chemistry professor who wanted to find an intelligent approach to tackling cancer. He postulated that if it wasn't possible to directly target the cancer itself, he could target what the cancer needed: the cell. Since cancer was an overgrowth of cells that simply could not stop dividing, the best way to kill the cancer was to kill the ability of the cell to divide.

Before a cell divides, splitting itself into two cells, it has to make a copy of its own DNA. Each cell needs its own set of identical genetic material. This DNA is formed from nucleotides, the "building blocks of life." The helix itself unwinds until the DNA is stretched out like a ladder. The rungs of this ladder break open, splitting up the DNA strand. The DNA looks like a zipper, the molecular bonds slowly coming apart. Each half of the DNA strand will eventually separate into different cells. New building blocks enter the opening DNA zipper, forming brand-new strands of DNA. The idea here is to make two identical copies of DNA from the one original. Like a zipper, the newly made DNA strands form the opposite tread. Each new strand fits perfectly into the original strand, forming two new functional zippers.

Because they are complements of each other, each half of the DNA strand carries all the genetic information for the cell. The DNA strands make a simple pattern of bases, each base intimately pairing with its one and only complement on the other side of the aisle. Adenine (A) invariably pairs with thymidine (T), and guanosine (G) pairs with cytidine (C). These building blocks are intimately connected; the correct binding of each of the nucleotides is necessary to the growing chain of DNA. A simple sheath of sugars and phosphates, all wrapping around each other and forming the double helix, connects them. Once two identical DNA strands are made, the cell can divide, forming two cells where there was only one.

Horwitz had a devilish plan to attack the replication of DNA. He devised the idea for a "trick nucleotide." Instead of a simple thymidine, one of the DNA building blocks, his thymidine would be altered. Once his trick nucleotide was incorporated into a cell, it would abruptly cut off the growing DNA strand. The cell couldn't divide; the cancer would be stopped. Horwitz worked tirelessly to make these trick nucleotides for all of the four bases, A, T, G, and C, that make up DNA.

Horwitz had a wife and growing family at home. Yet he was drawn to the lab, spending nights and weekends alone at the bench. He believed in his strategy. Perhaps this is why it was so devastating when it failed. He treated leukemic mice with the new family of drugs. Nothing happened. The tumors continued to grow; the runaway cells didn't even slow down.

It was 1964 and the world seemed to be coming apart at the seams. The Vietnam War was gaining momentum, violence was erupting at home as the Civil Rights Act was passed, and in labs all over the country, researchers were desperate to find a drug to treat cancer. Horwitz wrote up his failure. In his heart, he believed the drugs had a purpose to them. When describing the failure to his colleagues at Wayne State University, he would describe the drugs fondly as "a very interesting set of compounds that were waiting for the right disease."

Despite the tantalizing promise of these compounds, he didn't bother patenting them. Patenting drugs was not cheap.

Why bother wasting money on patenting failed drugs? He had already wasted precious resources on developing the drugs to begin with.

The failed compounds were archived, their records collecting a thick layer of dust in the Detroit lab. In one of those boxes was 3'-azido-3'-deoxythymidine, or AZT, one of many seemingly useless compounds. And there it would sit for two long decades.

While David Barry was mobilizing a team to develop the first drug effective against HIV at Burroughs Wellcome, a different sort of team was forming at Mary Lasker's National Cancer Institute (NCI) in Bethesda, Maryland. Wedged next to the National Institutes of Health (NIH), the NCI was a small campus of labs and offices surrounded by flowering dogwoods. Researchers could be found wandering the curving paths at all hours. With great forethought, the institute was located next to a hospital, giving researchers and clinicians the opportunity to mix. It also allowed blood samples to flow easily from hospital to lab bench.

Like Jerome Horwitz, who tinkered with cancer cellular machinery drugs in Michigan, Sam Broder was born and raised in Detroit. Broder started at NCI as a young clinical researcher in the early 1970s. At rapid speed, he moved up the ranks from clinical associate to head of the oncology department. By 1980, he was a vital member of the research force, poised to take on what he couldn't possibly know was coming: a new and unprecedented epidemic.

A molecular evolution was under way at NCI. Many of the basic molecular techniques that are commonplace today—sequencing, cloning, protein expression—were just beginning to emerge, all from the famed institute. It was a renaissance for molecular biology, at an institute that happened to comprise a group of unique young scientists filled with passion and poised to become the next generation of scientific leaders. Broder learned about the new disease, GRID (gay-related immune deficiency), in 1981. A young man who had recently visited Haiti had a strange set of symptoms, none of which fit together. Broder had never seen anything like it before. He said to a colleague as they discussed the unusual case, "I hope we never see anything like this again."

Despite the politicized nature of the burgeoning epidemic, to most scientists it was clear from the earliest days that the disease was not based on lifestyle. When a colleague at the NIH referred to the disease as relevant only to gay men, Anthony Fauci, head of the National Institute of Allergy and Infectious Diseases (NIAID), pointed out that the disease could be passed from mother to child and angrily responded, "What lifestyle did the fetus undertake to acquire the disease?" He presented clear evidence to critics, both scientists and journalists alike, that the disease was not based on a lifestyle or sexual orientation. He, like many other HIV researchers, was passionate about exposing the true nature of the epidemic.

When Robert Gallo, also at the NCI, announced that AIDS was caused by a retrovirus, HTLV-III, there was considerable excitement among policy makers, who were eager to calm the public's fears with a promised cure, but little enthusiasm from scientists. Standing next to Margaret Heckler, Ronald Reagan's secretary of Health and Human Services, that day in April 1984, Gallo felt increasingly uneasy about Margaret's words. He was exhausted, having rushed from an overnight flight from Italy to the packed, hot room. His pride stung from an article published only the day before in *The New York Times* giving credit for the discovery of the AIDS virus to the Pasteur Institute in Paris. He felt he had been unwittingly roped into doing a press conference. Now Margaret was making fantastic claims about the pursuit of new therapies. In fact, she announced that a cure would be available in only two years. Gallo felt hopeless to correct what he knew was obviously false. The press conference would continue to haunt him years later as it served to widen the growing rift between him and the French scientists.

Although the discovery of a cause for the rapidly growing disease should have been a relief, in fact, finding the agent behind AIDS only brought more worry.

For most clinicians, the news that AIDS was caused by a retrovirus meant one thing: There would be no quick and easy therapy. Retroviruses were notorious for their complex life cycles and the distinct lack of research on them. Retrovirologists had little experience in clinical drug development. This was not good news.

Not surprisingly, Gallo found that identifying partners interested in developing therapies for AIDS was not easy. Pharmaceutical companies in general were staying far away from the disease whose mechanism of transmission was far from certain, making it dangerous to work with, and whose market was still relatively small. At the end of 1984, while Gallo searched for pharmaceutical partners, there were fewer than 8,000 cases of AIDS in the United States. No one yet realized how soon that number would explode. Few companies wanted the risk of working on this new, dangerous disease. Companies were concerned that it would take heavy financial investment while profits were far from certain. The concern was part scientific since there was so little known about the infection. But in addition to the medical confusion surrounding the virus, reticence to work on the disease was also fueled by homophobia. AIDS was viewed as the "gay plague," a perception that dampened initiative from some pharmaceutical companies and research scientists.

In addition to homophobia, the disease itself inspired fear; some hospitals refused care to those who were HIV-positive, unwilling to admit patients whose condition frightened them. Firefighters banned the "kiss of life," afraid mouth-to-mouth resuscitation would transmit the deadly virus. Police officers in New York City began carrying masks and gloves for dealing with "suspected AIDS patients." The debate even spread to parents who worried about sending their children to school with children infected with the virus. The most famous of these cases was that of Ryan White, who, in 1985, as a thirteen-year-old hemophiliac with AIDS, was banned from school.

Despite the difficulties, the NCI decided to pour resources into research. They began pumping out massive quantities of the newly discovered virus, searching for a blood test that could be used to screen donated blood. Even though the institute encouraged its scientists to work on HIV, not everyone wanted to. Many believed that the politics of the disease were too complicated. Unlike other diseases, to work on HIV made researchers the subject of scrutiny from activist organizations that were not afraid to protest against companies and research centers they disagreed with. It also drew political lines. Activists believed that President Reagan's response to the AIDS crisis was poor (Reagan didn't mention AIDS until 1985), further politicizing the response to the epidemic. Many simply didn't want to waste time adding another project to their busy schedules.

Broder was not one of those scientists. From the time he saw the first HIV patient at the NIH in 1981, he found himself fascinated. As an oncologist, he couldn't help comparing the way the virus replicates to the way a tumor cell does. In both diseases the cell is overtaken, given signals and commands that are not within the realm of normal development. As the diseases progress, tumors metastasize, spreading from one part of the body to another. HIV similarly grows, from a single genetic variant to a diverse genetic circus able to infiltrate almost every tissue imaginable. Later, Broder would compare the development of therapy for HIV to that of cancer, saying, "Principles, drawn from the world of cancer, had significant implications for the development of antiretroviral agents, starting with AZT."

As he stood in his office, contemplating adding yet another project to his full roster, Broder stroked his thick, bushy, black mustache and adjusted his glasses. Now that the virus had been identified, the next step was obvious. He and Gallo needed a way to test for the virus. They needed to develop a system that could diagnose those with the disease. All the researchers knew that the national blood bank supply was in jeopardy if people donating blood didn't know they were infected. The supply of blood critical for saving lives could be massively reduced overnight. . . .

Broder's mind raced to the next step. Identifying the virus was necessary, but what about finding a treatment? How could they manipulate the science further to screen for drugs? There was only one place this work could be done: the NCI. Looking back, he would say, "It was clear that we needed a focused laboratory that was used to drug discovery and was willing to work with live AIDS virus. The only institute in the NIH that historically had focused heavily on new drugs is the cancer institute."

At Dave Barry's urging, in 1982 a small team at Burroughs Wellcome had begun pulling antiviral drugs off the shelves. What began as random shots in the dark changed radically in 1984 with the discovery that AIDS was caused by a retrovirus. Suddenly, the known biology of retroviruses could be manipulated.

The most obvious way to target a retrovirus, the Wellcome team reasoned, was to target the one unique cellular process that all retroviruses desperately needed: reverse transcription. Retroviruses, starting as simple strands of RNA, need to take over our cell machinery. Since they don't have cells of their own, they need ours. So they cleverly insert themselves into our DNA. The Wellcome team believed that this process would be the easiest to disrupt. Most of the team was focused on the known antiviral drugs lining the pharmaceutical company's shelves, but one woman had a radical idea.

Janet Rideout, an organic chemist working on the Wellcome team in North Carolina and interested in antibacterial drugs, had recently pulled out an old drug from the archives of the Michigan Cancer Foundation. It was the early 1980s and she was curious to see if these old drugs, developed in the 1960s, might be repurposed to fight bacteria. One drug in particular was effective against bacteria, azidothymidine, or AZT. Rideout had spent the last few years learning all about AZT, fascinated by its unique nature. As the team's focus shifted from bacteria to HIV, she didn't forget the unusual drug that attacked the cell. As she saw one by one the Wellcome teams' antiviral drugs fail, her interest in AZT grew.

Just as Horwitz had spurned throwing random compounds at cancer, Rideout also believed in a rational approach to the hunt for an AIDS drug. Her intimate knowledge of the mechanism behind AZT allowed her to see the big picture. The drug made sense: By terminating the growing chain of DNA that the virus created to insert itself into a host cell, it could stop the virus in its tracks. Of course, Rideout knew that lots of drugs make logical sense and yet still fail. The only way to test AZT's effectiveness would be to test it on actual HIV, not on other, related retroviruses. The company realized they would need collaborators; they simply didn't have a safe way to work directly with the deadly virus.

As fate would have it, at the same time that Wellcome began looking for partners, Sam Broder at NCI and Dani Bolognesi at Duke University came looking for them. Broder had developed an exciting new way to screen for HIV drugs. After failed attempts with other companies, Broder was desperate to find a pharmaceutical company that could contribute a library of potential drugs for screening. In addition, he wanted to find a company that was willing to make the substantial investment (at the time the average cost of bringing a new drug to market was $400 million) needed to turn a promising drug into a clinical trial.

When the HIV team at NCI met with the team at Wellcome, it seemed to be a match made in heaven. Years later, when the patents became more valuable, it would turn into an ugly brawl. For the moment, though, the stars were aligned, and the first AIDS drug, AZT, was born. Horwitz's failure had turned into a dramatic, unexpected success twenty years after the fact.

In the crosshairs of AZT were thousands of dying AIDS patients. "Trials are treatment," demanded AIDS activists, desperate for any kind of therapy. The idea that safety should be considered in clinical trials was secondary. The architects of AZT believed that if patients were going to die, they might as well die with some hope.

CHAPTER 6

The Days of Acting Up

AZT is not just a drug; it's the emblem of a disillusioned culture. Nowadays, when nearly two-thirds of adults born after 1980 are in favor of same-sex marriage, when gays serve openly in the military, when professional sports players identify themselves as gay and it is barely news, it is hard to imagine the stigma attached to sexual orientation and how powerful it remained in the 1980s and even 1990s. In the mid-1990s, when Timothy and Christian were first diagnosed with HIV, a rock musical was premiering in New York City. *Rent*, a modern take on the classic opera *La Bohème*, explores the lives of young artists in New York City in the late 1980s as influenced by AIDS. AZT plays its own role in the musical, forcing its users to carry beepers so that they can time their medication precisely. Perhaps even more telling, AZT is effective in the play only for bringing people together, not for treating the virus.

AZT remains a powerful cultural force, symbolic of a dark history. As it entered the public arena, many scientists believed that its arrival would herald an intense celebration. Finally, there was a drug to treat HIV. This did not happen.

AZT was created, screened, and supported by the U.S. government, so it might seem odd that a private company owns the full patent rights to the drug. The fact that a company owns these rights speaks to the desperation of the federal government, which had few partners in developing AIDS drugs and was in no position to contest patent rights. With few companies putting any resources into drugs capable of targeting the burgeoning epidemic, the Food and Drug Administration (FDA) felt it had a responsibility to rush the drug approval process at record speed. This rapid approval took place despite signs in early clinical trials that the drug was considerably toxic.

The first clinical trial of AZT was riddled with both problems and successes. Between February and June 1986, 282 people who were HIV-positive were given either AZT or a placebo. There was a dramatic decrease in mortality in those who received AZT, so dramatic that the regulatory agency couldn't justify withholding AZT from anyone. The FDA promptly eliminated the placebo in favor of the real thing. Not everyone benefited from the new drug; some patients who took it didn't experience any benefit. This is because the drug inhibits only one step in the virus's life cycle. For some viruses, this means drug resistance can build at a rapid pace. The virus mutates quickly, resulting in a viral enzyme that's able to distinguish between a valid building block and the trick nucleotide that is AZT. The genetics of both the virus and its subsequent mutation rate vary from person to person, and so some patients will develop resistance to the drug more quickly than others. Today, we're able to accommodate this with myriad drugs, each targeting different parts of the virus and its life cycle. But in the late 1980s, there was only one drug available.

At the same time that researchers were evaluating the effectiveness of AZT in preparation for its approval by the FDA, they were also measuring the toxicity of the drug. Since it was a drug developed for cancer but never used in people, it was difficult to find the right dose that balanced maximum effectiveness with limited toxicity. As AIDS was a deadly disease, most researchers believed it was worthwhile for patients to suffer from some toxicity in order to avoid death. Those measuring toxicity in the AZT collaborative working group in the late 1980s, therefore, reported only their results and gave general guidelines. This group couldn't recommend a standard dose. Prescribing physicians adjusted the dose up or down on an individual level in response to how well the patient tolerated the drug.

From the first human trial of AZT, it was apparent that the drug had serious side effects. A progressive increase in hemoglobin, the molecule that transports oxygen in the blood, was seen in patients on AZT. At the same time, platelets, which help the blood clot, took a dramatic dip downward. Thirty-one percent of AZT patients needed a red blood cell transfusion, compared to only eleven percent on a placebo. More worrisome was the evidence that the drug was suppressing the bone marrow. The precious stem cells in the bone marrow that eventually develop into the red cells of the blood were suppressed, so newly made red blood cells were suddenly in short supply. Bone marrow suppression is a typical side effect in chemotherapy. It causes headaches, dizziness, and fatigue. AZT became a kind of chemotherapy for HIV. Still, concerns over the toxicity of AZT were, for the most part, pushed aside as the drug entered wide-scale production. The results were published in *The New England Journal of Medicine* in 1987, four months after AZT's approval by the FDA. As Broder says, "A consensus on the safety and efficacy of AZT would have been impossible." After all, people were dying, they needed treatment, and there was no time for expanded safety and dosage testing before the drug was made available.

If there was something worse than the side effects of AZT, it was the price. The drug cost an astounding sum: $10,000 a year. It may not seem so much now, but AZT was the most expensive prescription drug in history. More than the side effects, this number outraged the HIV-positive community. It was hard to believe that a drug that had been developed by a university scientist in the 1960s and screened by government institutes could possibly cost so much. Analysts will explain that the high cost of drugs is needed to motivate companies to take risks and encourage creativity because drug development is an expensive and dangerous enterprise. Inevitably, promising drugs will require heavy investment only to fail. Some will even kill people. However, compared to other drugs developed by pharmaceutical companies, Burroughs Wellcome had made little investment in this drugs' advancement.

The pricing of AZT stirred both the gay and HIV-positive communities in unprecedented ways. The protests that broke out over AZT were fueled by more than unfair pricing and the drug's unavailability to those who were desperate for it. They also represented the ire of a community neglected. HIV was a disease that few newspapers wanted to cover, that no politicians wanted to talk about, and that many doctors refused to treat. Thousands of people marched in New York and Washington, DC, and at the Burroughs Wellcome U.S. headquarters in Burlingame, California.

By September 1989, the protests had gained considerable momentum. Seven protestors snuck into the New York Stock Exchange and chained themselves to the VIP balcony, displaying a banner that read SELL WELLCOME. The activists were relentless.

Despite the fact that those with much passion but little experience led the protests, they were enormously successful. Only days after the protest at the stock exchange, Burroughs dropped the price of AZT by several thousand dollars. The drug was still ridiculously expensive, but the fledgling AIDS activists had shown that they had political sway. The feeling of change was intoxicating. Protests over AZT incited the formation of many of the HIV advocacy groups we have today. In 1987, the AIDS Coalition to Unleash Power was formed, dedicated to protesting the tangled story of AZT and the government's role in it. Today that group, ACT UP, is still a powerful advocate for those with HIV.

The protests influenced more than HIV drug availability. They showed patient advocacy groups everywhere the power of protest. They changed the expectations of patients. No longer were patients content to wait for drugs to be approved. Instead, today they demand access to clinical trials. Patients now know that, with organization and passion, they can make the government fund research into neglected diseases.

Today, expanded and compassionate-use programs are popular for drugs in late-stage clinical trials. Diseases that affect only

a small number of people, such as SCID (severe combined immunodeficiency), also known as the boy-in-the-bubble disease,
receive considerable research funding. This support has been
achieved by grassroots patient advocacy groups that, like the
AIDS advocacy groups before them, wield the passion and organization of their members to bring new therapies to market.

The patent for AZT expired in 1995. Anyone can now make
or sell it. Jerome Horwitz, the original inventor of the clever drug
that tricks DNA, never shared in the enormous profits. In 1992,
Burroughs Wellcome, today GlaxoSmithKline, reported sales of
$400 million. AZT is still used today, typically as the first line of
defense in protecting infants from contracting HIV from their
mothers. When combined with other antiretroviral drugs and
given in much lower doses than those in the late 1980s, the drug
is effective against HIV.

For researchers, AZT represents the first success in the fight
to cure AIDS. It defeated what Sam Broder termed "therapeutic nihilism." Meaning it showed what many scientists simply
didn't believe was possible: We can treat a retrovirus. It was an
era when successful treatment for any virus was novel. The
breakthrough research that gave us AZT is the basis for all the
HIV drugs we have today. Robert Gallo, the American scientist
who shared in the discovery that HIV causes AIDS, believes
that AZT opened a new "window of opportunity" for the way
we treat the disease, saying that the drug spurred him to focus
on the human cell and its machinery when looking for new drug
targets. Similarly, Broder believes that the advent of AZT
changed the "cure or nothing" mentality that prevailed at the
time. While everyone wanted the whole loaf of bread, the

arrival of AZT brought the first slice. Many scientists quote Voltaire: "The perfect is the enemy of the good." The drugs we have today that target the interaction between viral enzymes and our cells are a product of this new kind of thinking. AZT was the drug that started it all.

CHAPTER 7

Recognizing a Global Pandemic

D avid Ho was born in the large city of Taichung, Taiwan. His parents struggled to support their young family, with his dad changing jobs every few months. When Ho was five years old, his father decided to change his luck for his wife and two children. He left them in Taiwan and moved to Los Angeles, confident that it would take only a year until he could save enough money to bring his family back together. Seven years later, the family was reunited when Ho, his mother, and younger sister moved to Los Angeles. He was used to living in a large city and excited to be living in America. He excelled in school, showing a particular aptitude for the sciences.

Ho studied physics at the California Institute of Technology (Caltech), then moved to the East Coast for medical school. His proud parents helped him pack up his belongings and make the cross-country move. At age twenty-six, he graduated from Harvard Medical School and immediately moved back to Los Angeles. It was 1981; Ho was chief resident at Cedars-Sinai hospital. A strange, new group of patients started emerging, patients presenting with unusual opportunistic infections indicating that their immune systems weren't functioning normally. As it would turn out, these were some of the earliest cases of AIDS in the United States. By chance, Ho saw four of the five first AIDS cases described. A report published by the CDC on June 5, 1981, documented this odd cluster of five homosexual men, all of whom developed a rare pneumonia. It wasn't known what caused their disease. When Ho recalls that time, he remembers how skewed his perspective was. "I was so focused scientifically on what was happening," he said, "I could have no idea this was the start of a global pandemic."

These early cases in his medical career influenced the course of Ho's professional life. After finishing his residency, he moved back to the East Coast to work at Massachusetts General Hospital in Boston. It seemed that HIV followed Ho. While he'd originally intended to work on another virus, herpes, he was drawn to HIV as the cases of the mysterious new infection rose in the wards of the hospital. Ho became, in his words, the "one and only person playing with samples from HIV-positive patients." He wasn't afraid of the risk of the new disease; he wanted to understand the science.

For his first research assignment in Boston, Ho was given a project on Kaposi's sarcoma (KS). KS is a cancer that leaves blotchy purple lesions all over the skin and mouth. It was the first opportunistic disease associated with AIDS. It's called opportunistic because it takes advantage of our immune system when it's down. Although KS is rare, it's frequently seen in people with AIDS, whose immune systems can't defend against the herpes virus that causes KS. As Ho studied the cancer prominent on the faces of those with AIDS, he couldn't help being reminded of another plague that caused spots on the face—"the speckled monster," or smallpox. The two diseases have radically different origins; AIDS is a retrovirus transmitted through bodily fluids, while smallpox is a large poxvirus that can travel airborne from person to person. Smallpox is one of the biggest killers in history. By some estimates, it is responsible for more deaths than all other infectious diseases combined.

While plagues differ in their details, symptoms, mortality rates, and whom they target, there is one common thread that runs through the history of all pestilences: stigma. This can be seen in such varied diseases as the Black Death in the four-teenth century, cholera in the nineteenth century, and AIDS in modern time. The writer Susan Sontag described this stigma perfectly as "the archaic idea of a tainted community that illness has judged."

Smallpox carried its own stigma. Although the virus is not transmitted sexually, the disease itself renders its victims repul-sive, covered in oozing sores. The bumps build over each other, covering the skin, and fill with a thick white fluid. For those spared by smallpox, the virus left them scarred and disfigured.

On May 14, 1796, Edward Jenner gave the first vaccine against smallpox. The forty-seven-year-old family doctor took a sample from a related virus and inoculated an eight-year-old boy, the son of a man who worked for Jenner. Although hard to believe today, two months and a second inoculation later, he tried to infect the boy with smallpox. It was the first vaccine in history. Thanks to that unethical experiment, the World Health Organization (WHO) was able to declare smallpox officially eradicated from the planet in 1980. The next year, 1981, a new disease, AIDS, would be discovered. David Ho hoped the lessons learned from Jenner and the first vaccine could be applied to HIV. Unfortunately, the same rules didn't apply. Where smallpox could be prevented by an inoculation with a similar virus, this approach hasn't worked with HIV because it is a rapidly mutating retrovirus. Where the plagues intersect is the effect they have on a community. Ho says, "If you walked in with an AIDS diagnosis you'd be dead within a matter of weeks. There was discrimination. Patients were not wanted by staff, friends, and family . . . discrimination pushed me to work."

By 1995, Ho's skyrocketing career had led him to New York City as director of the fledgling Aaron Diamond AIDS Research Center. He wrote an editorial that year for *The New England Journal of Medicine* titled, "Time to Hit HIV, Early and Hard." The article, which would become famous among HIV researchers, hypothesized that treating HIV early, using multiple HIV drugs, including drugs not yet approved by the FDA, could result in an "ablative therapy." He compared the strategy to the battle against tuberculosis and childhood leukemia, saying, "It was aggressive combination chemotherapy at the outset that led to cures. Optimistically, we can hope that such an approach will become possible in patients infected with HIV-1." The hope was that the virus could be eradicated in patients during acute infection, resulting in the desperately needed cure. If not a cure, at least the drugs could stem the virus before it took hold in the body.

Although mass killing of T cells comes later in infection, the virus still kills some cells in acute infection, particularly in our tissues. Once the T cells in our tissues are gone, they don't come back. This has prompted some researchers, like Ho, to advocate for early therapy. The logic for early intervention, Ho argues, is "infallible," explaining, "Even when a person is doing very well, virus is churning away, knocking off CD4 T cells. Why let this happen?" Despite the logic behind early therapy, Ho's theories were not universally accepted. Even now, Ho wishes the medical community had adopted his guidelines for early HIV therapy. He says, "It still bothers me that there's not greater consideration for the science. It's always 'show me the data' and yet this doesn't apply to other diseases."

Ho is referring to diseases like breast cancer. Despite a lack of evidence of the benefit of early screening and treatment, few argue against giving early, aggressive treatment to people whose mammograms have revealed tumors. Ho wishes that physicians would respect the science about what HIV is actively doing at this stage—making billions of copies of itself and crippling tissue immune cells—rather than insisting on large-scale clinical trial data. Ho uses a fitting analogy: "A man falling off a 100-story building feels fine when he passes the 50th story." So, Ho argues, does an acutely infected HIV-positive person. But this doesn't mean he doesn't need a safety net. While Ho's argument for early therapy wasn't universally accepted, and today still isn't, his work developing a new class of HIV drugs, which inhibit the protease enzyme, and his subsequent plan to eradicate HIV infection won him many accolades from both the medical and greater community. In 1996, he was named *Time*'s "Man of the

Year," and *Newsweek* ran an article enticingly titled "The End of AIDS?" featuring Ho's work. Everywhere, people with HIV celebrated the approaching end of a fatal era.

Bruce Walker came to medicine by a path littered with false starts, roadblocks, and confusion. He doesn't remember a time when he didn't love science. His dad, a geologist, inspired his fascination with the natural world. Walker describes him as a "workaholic." He recalls happy, precious Saturdays accompanying his dad to field sites. After one of these trips, when Walker was only eleven years old, his dad put a sample of pond water they had collected under a microscope. "It was teeming with life," Walker recalls, smiling. This was a turning point. Those drops of pond water held far more interest for him than all the rocks his dad studied. He felt a draw to biology in a way that geology couldn't touch.

Like David Ho's family, Walker's would move a continent away during his formative years. When he was a junior in high school, his dad got a grant to study red soils in North Africa. So they moved to Switzerland, which served as a jumping-off point for his dad to travel back and forth to field sites. Now enrolled in public school in a foreign country, Walker struggled academically and suddenly had to learn German. But though it was difficult, it brought the family together, and Walker fell in love with the alpine country.

Walker spent time painting houses and driving a fruit and vegetable truck around Switzerland during the years when most of his peers were starting college. But by the time he was finishing school in Colorado, now in his mid-twenties, his feelings were clear. He was desperate to go to medical school. His long road to medical school made his acceptance that much sweeter. Judging by the size of the small envelope he opened at his parents' home one afternoon, Walker was expecting a rejection letter. But as soon as he read the opening line, beginning with "Congratulations," he fell back on the couch, tears streaming down his face. "It was one of those situations where you realize how close emotions can be to one another," he recalls. It was actually going to happen. He was going to medical school. Through all the starts and stops of his undergraduate education, he had come through with a prize that truly mattered to him.

After graduation from Case Western Reserve medical school, Walker continued his medical training at Massachusetts General Hospital. As in the past, he was a little unsure where medicine was leading him and what he wanted to specialize in. The next spring, Walker began to notice an unusual group of patients, all presenting with very uncommon diseases, such as pneumocystis, a fungal infection of the lungs. Specialists from all over the hospital poured in as they tried to figure out what was wrong with these young men. No one seemed to have an answer. This struck Walker forcibly: There were still mysteries in medicine. Even in a hospital like Mass Gen, full of the best doctors in the country, a disease could baffle the foremost medical minds.

It was 1981 and the mysterious disease was still called GRID, or gay-related immune deficiency. Some called it the gay plague or gay cancer. No one knew what caused it or how it spread. Walker had been undecided about his future medical path, but now, with the rising crisis, he found himself needed.

. . .

In the early 1980s, both Bruce Walker and David Ho were receiving training in infectious disease at Massachusetts General
Hospital. Ho was a year ahead of Walker and already a superstar.
In 1984, Walker and Ho sat at grand rounds, a ritual meeting that
occurs in hospitals around the world, where medical problems
are discussed. This grand rounds was special, for it featured Robert Gallo, the codiscoverer of HIV. Gallo had just published his
paper in *Science* identifying the virus as the cause of AIDS. "It
was so exhilarating," Walker remembers. "Finally this thing that
had been killing people was identified as a virus."

Walker decided to pursue a research fellowship, even though
it meant giving up much of the patient interaction he cherished.
His experience in residency had changed his outlook on medicine. As he watched patient after patient die of AIDS in the
crowded infectious diseases wards, he realized it was research
that was needed. He felt helpless as a physician, unable to offer
his patients anything but supportive care. Walker had done
some research in college but hadn't liked it. This seemed like a
good time to give it another try.

He began what would become a long research career in Chip
Schooley's laboratory at Mass Gen. Walker was still something
of a naïf. His mentor told him he should research the cellular
immune responses to HIV. "I didn't know exactly what that
was," he remembers. He didn't have much experience in immunology. Schooley suggested that he start by measuring the
response of T cells to HIV, particularly the storm troopers of
the immune system: the killer T cells. The hope was that by

understanding how the immune system fought off the virus, they could understand why it was losing the battle.

Arriving in the lab, Walker was told not to talk to two researchers: Joe Sodroski and Craig Rosen. These two, Walker was told, "were doing really important work" and couldn't be distracted. If it sounds childish for lab members to be told not to speak to one another, imagine how it feels for a new researcher trying to gain a foothold in the field. As odd as it may sound, labs like this exist today. Even worse, many high-powered labs take the situation a step further, pitting their lab members against one another, each junior researcher in a battle to get the data first for the sole purpose of putting his or her name at the top of a manuscript. For Walker, it was tough. While he appreciated his mentors, he felt lost at sea. There was little supervision. He had no one to ask for help. The lab environment was tense. None of Walker's experiments were working. As a year of his fellowship passed without any appreciable results, Walker sank into feeling like a failure.

One Saturday morning, Walker was in the lab, and yet another experiment had failed. This isn't so unusual; all scientists have many failed experiments. A perfectly planned experiment faced with the chaos of reality often collapses. But, with persistence and experience, some will succeed. Part of maturing as a scientist is recognizing these moments, understanding when to pack it in and when to keep going. Walker, at the very beginning of his career, wasn't sure which to do. Morose, he sat looking over the botched data. Joe Sodroski, the researcher he wasn't supposed to disturb, came up to him. "What are you working on?" Sodroski asked, concerned about Walker's despondent attitude. Walker told Sodroski what he was trying to do and how nothing was working. Much to his surprise, Sodroski had the solution.

Using Sodroski's suggestions, Walker was able to get a unique model for HIV up and running. From their conversation, he devised an artificial system in which B cells taken from patients are engineered to express individual parts of HIV. He then measured the ability of killer T cells, the storm troopers, to respond specifically to each piece. With this scheme, he was able to dissect the components of the T cell response to the virus, teasing out which parts of HIV initiate a response. Taking the storm trooper cells from their first HIV patient, Walker's team was excited to finally measure how the body responded to HIV. What they found was dramatic. In people infected with HIV, their storm trooper T cells specifically target and kill HIV-infected cells. These killer cells knew who or what they were looking for. When storm trooper cells were taken from people not infected with the HIV virus, the cells did not know what they were looking for. This data was the first indication that the body can mount a specific response to HIV.

Walker's advisor knew this would be a big paper. He suggested they submit it to *Nature*, the field's preeminent academic journal. Walker had never written an academic paper before and found the process intimidating. Indeed, the paper came back with an ambiguous letter. It wasn't clear if the journal wanted to publish it. The reviewers wanted another experiment done. Specifically, they wanted HLA typing to be done on the patients being tested. HLA, which stands for human leukocyte antigen, is a cluster of genes that govern how our immune system functions. This cluster, located on chromosome 6 in our DNA, varies widely from person to person and gives us, as a species, an evolutionary advantage. With a wide variety of genes among us, we have a broad range of disease defenses. This makes it more likely that if a pandemic occurs and many die from a fateful disease, there will still be some of us who survive.

The journal wanted Walker to HLA-type the patient samples to be sure there wasn't a hidden genetic advantage influencing the strong response of the killer T cells to HIV. The problem was, no one who did HLA typing would touch samples from HIV-positive individuals. These technicians, like many people in the late 1980s, were afraid of working with HIV-positive samples at a time when it was still unclear how the virus was transmitted. So Walker decided to do it himself. He learned the procedure and then ran the HIV-positive samples. The data were, as Walker puts it, "equivocal." It seemed there was some pattern, but it wasn't clear whether this cluster of genes exerted influence. Walker sent the paper back to *Nature* with the new data.

While the paper was in review at the prestigious journal, Walker traveled to Washington, DC, for the third International AIDS Conference, in 1987.

Walker and his wife had just welcomed their first child, a son. Now he was also reveling in the excitement of sharing the new HIV data. His excitement was short-lived, however. As he sat through the opening talk by none other than Anthony Fauci, a prominent AIDS researcher and director of the National Institute of Allergy and Infectious Diseases, he was shocked to see his own data looking back at him. Walker, a self-described "lowly postdoc," was crushed. He assumed Fauci had been shown his data by a colleague looking for advice; it seemed Fauci had replicated their experiments.

Desperate, Walker called his mentor with the bad news. He told Schooley that T cell experiments very similar to theirs had been presented with no acknowledgment of their work. "Don't worry about it. *Nature* just accepted your paper," Schooley announced. In that moment, Walker's mood changed from depression to delight. His work had paid off.

As Walker finished his fellowship in Schooley's lab, his mentor sat him down to discuss his future. "Chip told me that one day I would be a famous immunologist, and I laughed. I was so sure he was wrong," he remembers, smiling widely. By the 1990s, Schooley was proven right. Walker was the director of the AIDS Research Center at Massachusetts General Hospital and Harvard Medical School in Boston.

Now that Walker had a way to compare storm troopers of various people infected with HIV, he wanted to see if the personal genetics of an individual's HLA affected the storm troopers' ability to target HIV. By chance, just as Walker was investigating the role of these killer T cells, he came across a few, rare individuals. These were people who, for reasons no one understood, controlled HIV without taking any drugs. Because they were such a rare group, they became known as elite controllers. Walker discovered that the secret behind why this group of people is special lies in the ability of their T cells. Unlike the defeated army of cells in most people with HIV, in elite controllers, the T cells survived and were able to coordinate an effective strategy of targeting and killing the virus. But how did the army maintain the ability to target and kill the virus? The answer seemed to lie in their genes.

Walker had discovered a fundamental truth about how the immune system responds to HIV and what is necessary to overcome the virus. Somehow, he and his colleagues had to figure out a way to replicate this preservation of the commanders and storm troopers of the immune system in people infected with HIV. But how could scientists re-create such a complex system of immune control without the genetic advantage?

Walker had a theory, intimately connected with the work David Ho was doing. He believed that early therapy might initiate the immune response seen in the elite controllers. He hypothesized that the use of antiviral drugs, if given early enough in HIV infection, would convert a typical person's storm trooper T cells into the elite squad that the HIV controllers maintained. But how could he find a patient given early therapy and then whose therapy was stopped? No physician could ethically stop therapy in a person infected with HIV. By chance, Walker was about to learn about a patient who fit these criteria—a young man given an unusual combination of antiviral drugs very early in infection: Christian Hahn, Berlin patient #1.

CHAPTER 8

From the One Percent

t the 1993 International Conference on AIDS in Berlin, the most depressing AIDS conference ever, Heiko Jessen was waiting to hear a talk from Bruce Walker. The year before, in 1992, Walker had come across an unusual patient, one who had become infected with HIV in 1978 in San Francisco. But he didn't know it at the time. Instead, the diagnosis was made later from blood products collected during a hepatitis B vaccine trial. Curiously, despite never having taken antiviral drugs, the man maintained healthy levels of T cells and did not progress to AIDS. Walker remained particularly interested in a special set of immune cells called the killer T cells, the storm troopers of the immune system. These cells are like trained killers. They have a finely tuned mechanism for detecting cells that are cancerous, infected with viruses, or are in some way damaged. Once they identify a cell that needs to be killed, they release cellular toxins, enzymes capable of tearing holes in the cell's membrane, eventually killing it.

The way these killer cells are able to identify a cancer cell or one that has been infected with HIV depends, in large part, on our own personal genetics. The commanders of the immune system, the helper T cells, have a receptor on their surface that is able to recognize specific pieces of the invader. But the receptors can't recognize these bits of the virus or invader by themselves. The virus must first be introduced to them. So bits of the invading virus's proteins, the antigens mentioned in chapter 3, are introduced to the receptors on the T cells by another cell, aptly named an antigen-presenting cell. This antigen-presenting cell encounters a virus and eats it up. It then takes the virus's antigens and hoists them up on its surface. Similar to how a conqueror would proudly display the head of its defeated enemy on a stake.

Seeing a head on a stake is enough to frighten anyone into action, especially a commander trained to defeat enemies of the immune system. But what if the commander sees a finger on a stake? It's not quite as scary. It is our individual genetics that determine which part of the virus will be displayed on the surface of the cell. While some people's display a head, a clear warning sign that something drastic must be done, others encountering the same invader show only a finger. The commander still mobilizes an attack against the invader but does not mobilize the full force of storm troopers, as it would in response to a head. (We biologists can seem macabre about body parts; forgive me.)

The response of the immune system is not determined only by the head or finger. Just as important is the stake. What holds the head is the human leukocyte antigen, but let's just call it the stake. This stake not only holds the head up but also determines which part of the invader is displayed. Depending on the genetics of your stake, you'll display a different part of your enemy to the immune system. This is key because which part is

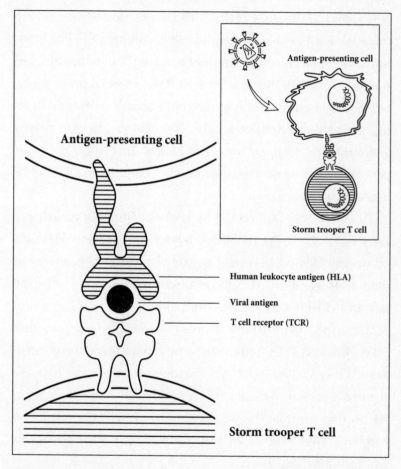

The immune system's version of a head on a stake. The response to intruders is measured by the interaction between three molecules. First, the human leukocyte antigen (HLA) on an antigen-presenting cell, a small piece of virus called an antigen, and the T cell receptor (TCR) on a T cell. How these three molecules fit together determines the strength of the immune system response.

displayed, the head or the finger, makes all the difference in how the immune system responds.

A critical interaction occurs between these three molecules: the receptor on the commander (TCR), the piece of the invader (antigen), and the stake (HLA). The interaction between these molecules can be seen in the illustration on page 81. The binding of the molecules determines the strength in which the immune system is mobilized. So if we have a special set of genes, when we present the virus to the commander's receptor it looks like a frightening head on a stake. Our commanders are alerted and our storm troopers are set to high alarm. Alternately, our genes can work against us, showing the receptor what looks like a nonthreatening finger.

Bruce Walker formed the basis of our thinking on killer T cells and how they're mobilized by our genes against HIV. It's no wonder that when he met a man who was HIV-positive but did not progress to AIDS, he immediately wondered what the person's killer T cells were doing and how HIV was being presented to the commander T cells. There's likely not another HIV physician at that time who would have made the connection. While Walker didn't quite understand what was happening in his patient, he knew it was significant. As he traveled to Berlin that summer, he was excited to present the results he had, and, even more so, he wanted to learn if other physicians had seen anything like it.

Also attending the conference was David Ho, who had moved to New York City in 1990. Working as an investigator at the New York University School of Medicine, he was desperate to begin clinical trials of new AIDS drugs at Rockefeller

University. He recognized the limitations of AZT, but he came
to Berlin hoping for good news.

It was a conference of failures. At the top of the list was
AZT. The drug was simply not keeping the virus in check; the
death rate for AIDS was climbing. Sadly, there were no other
options. Drugs from multiple other trials were presented, and
nothing worked. A trial called Concorde, which gave AZT to
patients early in infection, before symptoms appeared, was
worthless. Other, newer drugs, developed to target new parts
of the HIV life cycle, were a colossal failure. One study looking
at a new combination of drugs appeared to some researchers to
have arranged their statistics and study groups in order to show
a misleadingly positive result. The subterfuge didn't work. The
response was heated, as attendees accused the group of cheating. One researcher became so upset by the misleading data
that, during the discussion section, she angrily asked, "How
much is Roche paying you to say this?" By all accounts, the conference was ugly. Eight clinical trials, all of which had been
kicked off with high hopes, presented negative results. To top it
off, the death rate was the highest yet.

Jessen sat with his brother, sister, and Andrew at the conference. Although neither Jessen's brother nor his sister worked
with AIDS patients yet, they would both find their medical ca-
reers shaped by the virus in the years to come. His brother, Arne,
would even join Heiko as a physician in his practice. Jessen heard
the president of Germany mention AIDS for the first time and was
struck by the historical moment. He was even more stunned by
what followed. There were no new ideas, no hope. Andrew had
been diagnosed only two months previously, and Jessen had been

counting on finding some new drug with promise at the conference. He was sure there would be some inkling from an emerging clinical trial. Instead, they were left with nothing. He couldn't believe that even David Ho, the superstar in the HIV field, had nothing to offer. As he sat at the conference, he began to cry. There was no hope. Andrew would die.

In 1993, the Berlin AIDS conference crushed the hopes of those desperate for a new drug to treat HIV. The drug they were waiting for was still two years away from being approved by the FDA for use. That drug, saquinavir, was designed from the crystal structure of HIV's protease enzyme. While, in 1989, everyone thought Merck was the first to deduce the crystal structure of the viral enzyme, molecular virologists at Roche knew that the structure was actually quite different. They developed the drug RO3I-8959, or saquinavir, based on their modeling of the target.

What no one expected was that the combination of AZT, which targets the reverse transcriptase enzyme, and saquinavir, which targets the protease enzyme, would result in a powerful synergy. By combining AZT and saquinavir, the concentration of each drug increases exponentially inside the cell, making both drugs better at attacking the virus's replication. This combination therapy, called highly active antiretroviral therapy (HAART), would prove itself capable of knocking out the virus to undetectable levels in the blood (but not the reservoir) in the years ahead.

Fast-forward three years to the AIDS conference in July 1996, where David Ho showed a pivotal slide to the eager audience. The slide presented evidence for what seems obvious now:

The average person infected with HIV makes billions of copies of the virus daily. The data marked a turning point. It's hard to imagine now, but before this conference, physicians were divided on whether the infection even necessitated treatment. After this conference, the picture was clear: Patients needed to be put on antiviral drugs. Even better, new drugs were now available. In 1997, two papers in *The New England Journal of Medicine* confirmed what had been reported at the 1996 conference. Combination therapy, or HAART, reduces death by 60–80 percent. The 1996 AIDS conference was the opposite of the 1993 conference in Berlin. It was full of optimism. The word *cure* hovered around the 1996 conference, unspoken by the attendees but present in everyone's thoughts. The rise of a new class of protease inhibitors was a game changer. Hidden within the optimism of the 1996 conference was the hope that this combination therapy would be powerful enough to end the epidemic. Researchers hoped the drugs could purge the virus from the body. At the forefront of this hope was David Ho, the researcher pioneering the use of protease inhibitors and sparking headlines heralding the end of the AIDS era. But, despite the strides the new combination therapy made, it wasn't a cure. Not yet.

CHAPTER 9

But, Doctor, I Don't Feel Sick

I n 1996, just as today, experts had not agreed upon the best time for a patient to start treatment with antiviral drugs. Although some scientists hypothesized that early therapy could confer benefits, there was no hard evidence. Many patients had difficulty tolerating the drugs, which caused a range of side effects from psychiatric disturbances to gastrointestinal distress to fat redistribution. At St. Clare's Hospital in Hell's Kitchen in New York City, the AIDS ward was packed with young men and women suffering from all these side effects. One young man craved only ice, unable to tolerate solid food. Another lived in a state of dementia, confused and hallucinating. Almost everyone exhibited the sunken cheeks that marked a person, as prominent as a scarlet letter, as HIV-positive. Because of this range of side effects, physicians could not depend on universal guidelines, and instead had to judge for themselves whether to start therapy right away or wait until the effects of the virus could be measured in the patient.

HIV infects the majority of people as a single virus. It enters a cell and begins to invade. Its victims are T cells, a type of white blood cell. No matter how the infection occurred, the invasion begins primarily in the intestines and rectum. We tend to think of HIV as a disease in the blood, since the majority of research and testing has focused on this part of the body. Actually, most viral replication takes place in the intestines and rectum, where a dense network of white blood cells, including T cells, reside. The gut contains the vast majority of the body's immune system, more than 70 percent of all T cells reside there, not in our blood. The gut is the battleground for HIV and a broad range of infections. Following ingestion, sexual transmission, and even intravenous transmission, the gut is the first place HIV takes on the immune system. It isn't clear why this is so in the case of non-anal sex and intravenous infection, but the explanation may lie in our immunological past.

HIV is able to break into T cells because of the proteins the cell holds on its surface. HIV needs two proteins to sneak into the cell. The first is CD4. T cells with the protein CD4 are the commanders of the immune system. They coordinate the attack and send in the killer T cells, the storm troopers of the immune system, to clear the virus. The fact that HIV first identifies and takes out the commanders is a shrewd tactical strategy, for without its commanders, the immune system can no longer coordinate its attack on HIV.

But HIV needs more than just CD4 to enter T cells. The presence of a second protein, CCR5, is critical. The vast majority of viruses need CCR5 to enter our cells. This human protein serves no real purpose in our bodies. Like our appendix, its presence or absence doesn't seem to affect our health. The CCR5 protein lies next to CD4 on the cell surface. Like opening a locked door, HIV's contact with CD4 and CCR5 acts as a key fitting a lock. As shown in the illustration below, the virus first forms a tight bond with the CD4 protein on the surface of the cell. Then it also grasps CCR5.

The part of HIV that is the key to this unusual lock is an ingenious tool. Each particle of HIV is covered in little spikes, which are the envelope protein of the virus. Each of the spikes is needed to get into the T cell. The spikes themselves are split into two distinct units: gp120 and gp41. The gp120 unit is located on the tip of the spike, while gp41 is at the bottom. As the virus

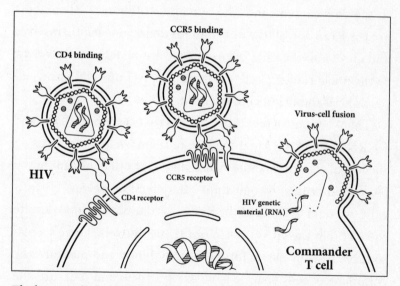

The key to unlocking a T cell. HIV's envelope protein first makes contact with the CD4 receptor on the surface of the T cell. Once it has bound to this receptor, it then binds a co-receptor, CCR5. This interaction causes the envelope protein to fold in on itself, bringing the virus close to the cell. Once virus and human touch, the membranes fuse, allowing the virus to enter the cell.

approaches the cell, either free-floating in our blood or trapped inside the tissue of our gut, the top part of the spike, gp120, binds to CD4. When that initial contact is made, the virus is drawn close to its cell victim. The bottom of the spike is held in just the right spot as it makes contact with CCR5, the close neighbor of CD4.

Once HIV holds both CD4 and CCR5, the bottom part of its envelope spike begins to fold in on itself. This folding draws the virus close to the cell so that the membranes of the two are flush against each other. Like two drops of water running down a window, the virus and the cell finally meet, and their membranes mesh. The two drops become one. The virus is now able to spew its contents into the human cell. Those viral contents are the RNA and all the enzymes needed to unpack themselves and move into the cell. Once inside, the viral RNA travels right to the nucleus of the cell, ready to take over our DNA and cell machinery and begin copying itself.

There are only a few types of cells that carry both CD4 and CCR5. While HIV can infect any of them, we tend to focus on the loss of the commanders since they are both abundant in the blood and critical to our ability to defend ourselves. Historically, we tend to think of the disease as one of the blood despite the fact that the virus lives in our tissues as well. There's a reason for this. It's easier for us to draw blood and measure the commander cells than it is to excise tissue and measure the other cell populations that carry CD4 and CCR5.

The viral envelope scans the cells that surround it, detecting the CD4 and CCR5 proteins that poke their heads out of our cells. Once it detects these proteins, it binds to them like a magnet, gaining entry into the cell. Apart from commanders, HIV attacks cells called macrophages, white blood cells that ingest invading pathogens. Macrophages are sometimes called the garbage disposal unit of the body. HIV doesn't kill macrophages directly. Instead, it keeps the cells alive, even going so far as to change the way our body communicates with these little garbage disposal units. The strategy is ingenious, for macrophages can travel anywhere in the body, carrying the virus with them. HIV is most famous, though, for its destruction of T cells, the cells that regulate our immune system.

Commander T cells are perfectly round and covered in fuzzy spirals of the CD4 protein. Again, these commanders of the immune system don't directly kill cells infected by virus or bacteria. Instead, they coordinate the response to the infection, activating the storm troopers, or killer T cells, that, as the name implies, directly kill cells infected by the virus. The commanders also activate the B cells, which, like a bomber squad, drop antibodies on the virus, mangling it and making it difficult to infect new cells. In each cubic millimeter of blood, about the size of a raindrop, the average person maintains a healthy level of 500 to 1,500 commanders, but at the height of HIV destruction, this number can drop to zero. On the road to zero, HIV kills a lot of T cells. HIV acts like a trained assassin, killing the commanders that the rest of the military depends on.

But here's the rub. During acute infection, when the newly invading army, HIV, is rallying its troops, there are actually few symptoms, and they are all mild. Patients experience flu-like symptoms common to viral infections, such as fever, achiness, and fatigue. This stage is not about killing T cells, although many

will perish, mostly in the tissues. These losses are minor compared to the massacre about to occur in the weeks ahead. The
commander cells aren't noticing anything unusual going on.
During this time, the virus is ramping up, making as much of
itself as it can, on average 10 billion copies a day. Two critical
events are linked: a peak of virus occurs in the body just before
the number of commanders plummets. Externally, the person
looks healthy, and likely feels healthy, but inside, their immune
system is crumbling.

Within the first few weeks of infection, the virus kills cells
locked away in our tissues. These cells, comprising both commander T cells and macrophages are, unlike in the blood, packed
closely together. They are the perfect first victims for the virus.
We don't typically measure the cells that line our gut and genital
tissue. From a medical perspective, we're not likely to notice that
they're gone. Once the virus kills these cells, the cells don't come
back. Even after decades of antiviral therapy, we can't replace this
precious hoard of commanders. As the virus makes more and
more of itself, it seeps into our blood. There is no magic switch
we know of that occurs to mark the end of acute infection and
the beginning of chronic infection, which can lead to AIDS. Instead, the virus seems to reach a critical mass, then the blood is
filled with billions of copies of the virus. The immune system
responds but is overwhelmed. The destruction begins. While the
commanders are the first victims, their decimation takes time.
As cells are killed, the immune system is left defenseless. And
just like that, a person progresses from HIV to AIDS.

So when should a doctor begin therapy? Early on, to prevent
damage to the immune system caused by the virus? Or later,

after the effects of the virus can be measured in the blood and the need for medication is obvious? The drugs prescribed remain the same whenever therapy is started. However, the general wisdom around 1996 was that, once started, HIV medications should never be stopped. Stopping the drugs could open up the virus to mutation. And once mutated, the virus could develop resistance to the antiviral drugs.

Doctors hate to start patients on therapy when there is no sign of disease. With no evidence that early therapy held any benefit, there was simply no reason to start therapy before symptoms developed. In fact, it was risky to start therapy too soon because patients might be tempted to stop the medication and then develop drug resistance. After all, it's asking a lot of someone to take medication when they don't even feel sick—particularly when that medication isn't easy to take and causes awful side effects. So while in 1996 we finally had effective, new drugs for HIV, what we didn't have was the instruction manual. We were about to get a sense of just how much the burgeoning science of personal genetics could help us in the search for a cure for HIV.

CHAPTER 10

The Delta 32 Mutation

The year that definitively shaped the Berlin patients' cure was 1996. The new antiviral drugs available that year represented a watershed in HIV research. But something else was discovered that year. An odd finding about the effect individual genetics have on controlling HIV that would end up being just as, if not more, important. It was all about the gene known as CCR5. Or rather, it was all about a particular mutation of that gene: delta 32 (Δ32).

In the early 1990s, a small group of gay men in New York City realized that despite having risky sex multiple times with HIV-positive partners, they remained HIV-negative. Some of these men wondered why they remained disease-free, and sought an explanation. Eventually, twenty-five of them made their way to the Aaron Diamond AIDS Research Center on the east side of Manhattan in New York City, where David Ho was the research director. The center is the world's largest private research center dedicated solely to HIV. This group of men became known as the

EUs, for "exposed uninfected," and were established as a patient cohort at the research center.

In 1996, this group of researchers at the New York City research center published a landmark paper. They had uncovered the reason why these EUs remained uninfected despite their risky behavior. The men had a mutation in their CCR5 gene that causes 32 pieces of the gene to be missing. This came to be known as the Δ32 mutation.

The CCR5 gene encodes the CCR5 protein, and scientists often call it a good-for-nothing gene because its role in the body isn't essential. CCR5 stands for chemokine receptor type 5. Chemokine receptors sit on the surface of cells and interact with a small family of chemotactic cytokines, collectively called chemokines. Chemokines are like magnets in the body, directing proteins where they need to go. It's believed that CCR5 directs the movement of proteins around the body in response to chemical signals. Whatever role CCR5 plays, it doesn't seem to be a very important one. People who have the Δ32 mutation in this gene don't express the protein in their body and it doesn't seem to affect their health. If you have the mutation, you probably don't even know it.

So, although the CCR5 protein seems to play no significant role for us, having it makes us vulnerable to the HIV virus, which uses it to invade our cells. And although the Δ32 mutation also seems to have no purpose, nor causes harm, it protects against HIV. Without a functional CCR5 on the surface of the T cell, HIV isn't able to get inside. It can't infect a single cell. When the virus can't enter the cell, it's slowly filtered out of the body, unable to hurt anyone. Like an out-of-luck gate-crasher, the virus is locked out.

The good news is that the Δ32 mutation is surprisingly common. It's found in about 1 percent of Europeans. People who are homozygous for this mutation (meaning that both copies of their CCR5 gene have the deletion) will be resistant to HIV infection throughout their lives. There are also those who are heterozygous for the mutation, meaning they have one mangled CCR5 gene and one normal copy. They express lower than normal levels of CCR5 on the surface of their cells and there is some evidence that even this confers an advantage, slowing the progression to AIDS.

Slowly, researchers around the world were putting together the pieces of how the virus could be controlled. But the question remained. How to turn this knowledge into a therapy capable of saving lives?

Gero Hütter read the 1996 papers on the Δ32 mutation and HIV with interest. Hütter was in his third year of medical school at Humboldt University of Berlin, Germany. He wasn't particularly interested in infectious disease or HIV. He was focused on hematology and oncology. He was in his early twenties and spent almost all his time studying. He didn't enjoy being a student. He had struggled in school before the doctor shortage in Germany pushed him into pursuing medicine. He was already dreaming of what would happen when he finished his medical and research training. He knew he wanted to stay in Berlin. He loved the city and the possibilities it offered for research. It was a competitive environment for academics, and the chances were slim that he would become a faculty member at one of the major medical schools in Berlin. But Hütter knew this was what he wanted, and he was willing to work for it. He daydreamed of working with cancer patients at the Charité hospital, performing exciting research, maybe even curing cancer. HIV was very far from his mind. Nonetheless, when Hütter read the papers detailing how the Δ32 mutation was able to confer protection against HIV, he was struck by the enormity of the finding.

Hütter sat in the medical school library with the journal in his hands and looked out the window at the icy sleet raining down against the glass. *It's so simple*, he thought. *One mutation and HIV can be stopped*. He sat back in his chair. Hütter believed that with such a striking finding and the research coming from David Ho's lab in New York City it wouldn't be long before the disease was cured. It seemed obvious that this was a special moment in the history of HIV. Indeed, magazines and newspapers were brazenly declaring the end of AIDS. It seemed that the research findings he held in his hand were likely to be a part of that ending. As he put the copy of *Nature* back on the shelf, he had little idea how much that paper would mean to him and the care he would be providing to Timothy Brown in the not so distant future.

Meanwhile, Timothy Brown and Christian Hahn grappled with their new HIV diagnoses. Timothy struggled with the side effects of AZT, while Christian was overwhelmed by his complicated drug schedule. Each would face a moment when he believed he was close to death. It turned out that they were not.

CHAPTER 11

Calling All Elite Controllers

n 1995, Bruce Walker was a successful physician and researcher at one of the top hospitals in the country. That year, he met a man named Bob Massie, whose genes would take Walker's lab in a new direction. In his poignant memoir, *A Song in the Night: A Memoir of Resilience*, Massie recalls Walker's initial doubt that the healthy man standing in front of him was, in fact, HIV-positive. Massie had been infected by a blood transfusion to treat his hemophilia when he was twenty-two. For the past seventeen years, he had somehow remained healthy, despite the fact that he hadn't taken any antiviral medications. Massie, who was engaged and wanted to be able to give his fiancée an answer to his medical mystery, was hoping Walker would be able to figure out what was going on inside his body. Walker confirmed that Massie was, in fact, HIV-positive, by an antibody test. It wasn't clear how he was able to control the virus.

Walker was still interested in how killer T cells, the storm troopers of the immune system, defend against HIV infection. Unlucky for us, HIV takes out the commander T cells, first thing, which means we lose the cells we need to coordinate our immune system. It's also bad luck for the virus, of course, which just wants to make more of itself. With us dead, it can no longer do that. The fact that the virus kills us is a sign that, evolutionarily, we haven't coexisted with it very long. With a little more time, we'd have found a better way for us to survive together. Successful viruses don't kill their hosts; they find a way to flourish within them.

The world is filled with little creatures that favorably live with larger ones. One hundred trillion microorganisms live peacefully within our gut. The white-gray patches on whales are actually small creatures called barnacles, which live happily with the large mammals, as much as a half ton of the animals on a single humpback. In some ways, humans have more in common with HIV than whales do with barnacles. Like HIV, we have a flawed relationship with what we need to live. Like the virus that destroys the cells it needs to survive, we often destroy, by activities such as deforestation and pollution, the very habitat we depend on for our survival. HIV takes out our commander T cells because they express CD4, the protein HIV needs to enter our cells. With the commanders gone, the immune system can't mount an effective attack: the storm troopers don't know where they're supposed to go and whom they need to kill. Without the commanders, the bombers aren't given the signals they need to drop antibodies that can bind up the virus. Without the commanders, the body is so shaken that it can't remember if it's seen the virus before. Even more insidious, HIV takes out the commanders during the asymptomatic stage of infection, before a person even knows he's infected, when he still feels healthy.

What Walker noticed right away when he looked at Massie's

blood was that, surprisingly, he still had his commander T cells. It wasn't simply the presence of the commanders that was unusual; it was the fact that these cells were also HIV-specific. The commanders can specifically recognize that a cell is infected with HIV and mount a vigorous response. The army of T cells in Bob Massie was larger than any Walker had ever seen in an HIV-infected person.

By chance, Walker had stumbled onto a patient able to control HIV through the very mechanism on which he was an established expert. It was clear to Walker that he had to figure out how Massie's commander T cells were preserved. From his early work on HIV and the immune system, he already had a hint. He suspected that the HLA, the genes that govern our immune system, were behind this remarkable control of HIV. The only way to know if the genes were the underlying cause was to find other people like Massie, who controlled HIV in a similar way.

Walker was giving a talk in New York City six years later when the tide turned. The talk was an update on the science of HIV and AIDS, which he was presenting to three hundred physicians and nurses who saw large numbers of HIV-positive people. Walker casually mentioned Massie, who was often on his mind. He asked the clinicians if any of them had seen a case like it. Over half the audience raised their hands. "I must have audibly gasped," Walker remembers. Here was the answer. If Walker reached out to enough HIV clinics, he could compare the HLA genes among these elite controllers. If they all had some particular gene in common, then perhaps there would be a way to get that gene into HIV patients who didn't have it.

There was the problem of raising enough money to do even

initial experiments. Walker believed there was a genetic tie among elite controllers and that it lay in the HLA genes. But he couldn't specifically say how that connection worked. No government agency, the typical partner in such research, wanted to fund an experiment that didn't know what it was looking for. During this frustrating time, Walker had breakfast with Mark and Lisa Schwartz. Mark, an investment banker at Goldman Sachs, and his wife, Lisa, an organic farmer and cheese maker, were funding an effort by Harvard to train African scientists and physicians to work on the HIV crisis. Mark asked Walker what else he was working on. Walker told him about the elite controller project and his difficulty in getting anyone to fund it. Mark and Lisa immediately understood the logic behind the project. That day, they committed $2.5 million dollars to collect samples from elite controllers. Walker then began making phone calls to collaborators around the world.

This approach to treating HIV, through the individual genetics of those who control the virus, was part of a growing trend. The promise of personalized medicine is that a patient's genes can inform our understanding of disease, indicate appropriate therapy, and determine likely side effects. As the cost of sequencing a patient's genes has dropped, our understanding of the intersection between disease and genetics has grown. Currently in clinical trials, we have investigational new drugs that repair the mutant gene responsible for cystic fibrosis. We have drugs capable of targeting specific proteins involved in cancer cell proliferation, as uncovered by genetics studies. The gene therapy field once struggled under the weight of seemingly insurmountable safety issues, suffering a major setback when an eighteen-year-old died in 1999 at the University of Pennsylvania. This resulted in the FDA's suspension of numerous clinical trials. But today the field is in a renaissance of sorts, with positive data being reported from such wide-ranging clinical trials as inherited blindness, Parkinson's disease, and inherited blood disorders. The current challenge in genetics-based medicine is that we simply have too much data. It's difficult to sort through which relationships are important and which are happenstances. For HIV, you would want to find a group of people who all had the same genetically powered machinery that enabled them to control HIV. Researchers had learned of the connection between the mutation known as Δ32 and HIV resistance. But Walker's genetics study of elite controllers was about to uncover a new way to keep HIV under control.

Dan Forich lays spread on the hospital gurney, feeling cold and nervous in his thin hospital gown. He flew from San Francisco to Boston to have the procedure: a routine upper and lower endoscopy, in which a thin probe attached to a camera is run down his throat and then another is run up his anus to obtain intestinal tissue samples. It's a common procedure used to check for polyps and the beginning of intestinal cancers. Forich smiles and shakes his head when asked if he has any questions, but deep down he's concerned about the anesthesia and what will happen if polyps are found. Mostly, Forich just feels hungry, for the preparation before the surgery meant he has not eaten in over twelve hours, instead having flushed his system with a particularly disgusting liquid he was told was needed to clean his intestines.

Forich has lived with HIV for more than two decades. He's watched close friends succumb to the disease, and most heartbreaking, he lost his boyfriend, who died from AIDS-related complications. Yet Forich remains healthy and, importantly, has never taken any antiviral drugs.

He is not alone. It is estimated that 1 in 300 Americans and 1 in 100 Europeans diagnosed with HIV will be able to control the virus without medications. In total, about 1 percent of the HIV-positive population doesn't need to take medication. Within this special group of people who can control HIV are subsets. Elite controllers essentially have undetectable virus in their blood: less than 50 copies per milliliter. Viremic controllers, on the other hand, are those with detectable virus, 50–2,000 copies per milliliter. Both groups are able to control the virus without therapy, although the long-term prognosis is better for elite controllers. Because they carry so little virus, it is nearly impossible for controllers to transmit it. However, the picture isn't all rosy; viremic controllers will sometimes, unexplainably, after decades of controlling the virus, suddenly slip toward AIDS.

It's important to remember that even if an elite controller has undetectable levels of virus in the blood, this doesn't mean there isn't virus hiding in other tissues. A special tissue called the gut-associated lymphoid tissue, or GALT, which lines the intestines, hosts the vast majority of the body's immune system. Unlike the blood, where immune cells are free-floating, the GALT forms a dense network of disease-fighting cells.

Because so much of the immune system is concentrated in the gut, this is where the body forms its first line of defense against outsiders. Mucosa-associated tissue lines these battlegrounds at the nose, throat, tonsils, intestine, and urogenital tract. In order to defend the body, the GALT hosts a large number of lymphocytes, which can identify the intruders and mount an attack.

While for most diseases it's advantageous for all the immune cells to be waiting to pounce, HIV is not so easily defeated. For HIV, this tissue isn't a threat, it's a welcome mat. Up to 90 percent of intestinal cells express CD4. In addition, lymphocytes in the gut express so much CCR5 that researchers originally thought the CCR5 receptor was unique to the gut. It's the perfect place for HIV to infect and take over, ramping up billions of copies of itself that can be spread throughout the body. In addition, the gut is the ideal hiding place for the virus. It can remain dormant in the gut for decades, long after antiviral drugs have wiped out the virus in the blood. For reasons we don't understand, the virus wakes up and returns to its full force. So the road to defeating HIV has to lead through the gut. Without considering what's happening in this key part of our immunity, we are doomed to continue harboring the virus and never being able to clear it. This is why HIV researchers ask for so much from HIV controllers and from the Berlin patients. Researchers need to know how they are able to keep the virus in check not only in the blood but in tissues, too.

Besides their ability to control HIV without medications, perhaps what is most remarkable about HIV controllers is their generosity in aiding HIV research. Hundreds of HIV controllers like Dan Forich undergo invasive surgery and long-term testing to help in the fight against the AIDS epidemic, without any direct benefit to themselves. Dan talks about this as he lies on a gurney, about to go into the invasive procedure. When asked why he's doing it, his answer becomes twisted, turning into a thank-you to the researchers, seemingly oblivious to his own contribution.

When asked what he thinks about the science behind his incredible control of HIV, Dan says, "I don't know. I guess I'm just lucky." He has been contributing to research for more than a decade. This is the second time he's flown across the country to undergo a voluntary and uncomfortable procedure. Yet, despite all these years of exposure to cutting-edge research, no researcher has ever sat down with him and explained how it is that he's lived with HIV for so long without developing AIDS. Somehow, science has been left out of informed consent. Where Jessen spends time explaining the biology of HIV to his patients, few researchers are able to spend this kind of time with their study subjects. Patients may understand the risk behind the procedure or therapy they receive, but it's unlikely they've discussed the science.

People like Dan are able to control HIV for such a long time without therapy because of their genetics. We understand this because of the gamble Mark and Lisa Schwartz made on Bruce Walker's research in 2002. Walker's wild hypothesis turned out to be right. Elite controllers have special genes that reside in chromosome 6 and encode the HLA, the human leukocyte antigen. HLAs are incredibly diverse in humans. Our individual HLA encodes a set of proteins that are then displayed on the surface of every single one of our cells. These HLA proteins are the equivalent of a secret handshake. If a cell has them, the immune system knows the cell is human. If the cell doesn't, it will be tagged as foreign and destroyed. This is why, when someone is getting a tissue transplant, whether it is liver cells or stem cells, the HLAs between the donor and the recipient must match. That way the donor cells are recognizable when they get into the body and it's less likely the transplant will be rejected by the body.

These proteins play a key role in HIV infection as well. When the virus enters the body, it's eaten up by antigen-presenting cells, or APCs. The APCs digest the viral proteins and then load bits of virus, the antigens, onto the HLA proteins that lie on the cell's surface. The head on a stake. They then bring the antigens to T cells. Like pieces of a puzzle, the T cell, viral proteins, and APC fit together. The signal the T cells receive from the APC determines the kind of response the immune system will mount. It turns out that, for HIV controllers, the message is loud and clear. The antigens that controllers display are very different from those of people who progress to AIDS. The virus acts as a double agent in HIV controllers, secretly signaling to the T cell that, yes, the threat is real and the immune system has to give it everything it's got. How HIV controllers mobilize the commander and storm trooper T cells is shown in the illustration on page 108.

It's not just that HIV controllers tend to have similar HLA genes, although they do. Specific HLA-B genes, such as B*57 and B*27, are found at disproportionately high levels in HIV controllers. This is similar to macaque monkeys; animals that

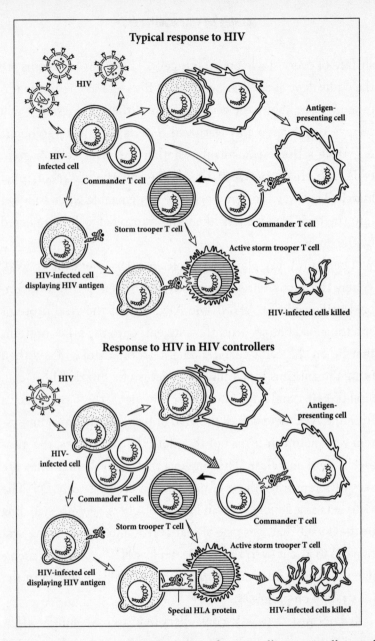

How controllers defeat HIV. The virus infects T cells in controllers and those that progress to HIV in the same way. Antigen-presenting cells sense the invader and swallow the virus. Controllers stimulate the production of HIV-specific commander T cells and storm trooper T cells, while progressors can't naturally mount as strong an immune response.

have the HLA Mamu A*01 gene are more likely to control SIV, the primate counterpart of HIV.

But it turns out that it's not the gene that's truly important. The real difference in HIV controllers lies in the individual amino acids that make up the groove on the HLA protein. The majority of HIV controllers have specific amino acids in this one region on the surface of the APC. Only a few altered letters of DNA make all the difference in whether or not a person can control HIV. So the difference between a person whose body can naturally control HIV and one who cannot goes beyond genetics. The real secret lies in a small section of the HLA-B gene encoding three amino acids. People who have these three amino acids—serine at position 97, methionine at position 95, and tryptophan at position 94—are likely to naturally control HIV through a coordinated immune system attack. Their bodies are able to present a special part of the virus to the T cell, unleashing the full force of the immune system onto HIV.

Sitting on the surface of all the cells in Dan Forich's body is this handful of amino acids. They don't give him an advantage in any other disease. In fact, they might make him more susceptible to certain autoimmune disorders such as psoriasis. But what these special amino acids do on Dan's behalf far outweighs any potential danger they might present. They protect him from AIDS.

A person without this genetic blessing has no inborn mechanism to control the virus. Now that scientists understood the basis of elite control of HIV, how could they translate it into a therapy for those without the special genes?

CHAPTER 12

Treatment in Hiding

As far as love stories go, Heiko Jessen and Andrew's story was passionate. Their four years together had forged the kind of love that even an HIV diagnosis couldn't break. As Jessen's fears became ratcheted up by the lack of any acceptable therapy capable of treating the virus, his desperation turned into action. He reached out to every scientist to whom he had even the slightest connection. He was desperate for new drugs, clinical trials, anything that might save Andrew. He still perceived the illness as Andrew's alone. He hadn't tested himself or even admitted his own stark risk for the disease he regularly counseled others to treat early.

One of the phone calls he made was to Robert Gallo, the prominent American scientist, the codiscoverer of HIV. Jessen had spent a few months training with Gallo's group at the NIH. He had seen Gallo speak many times. At these lectures, he had been his usual engaging self, asking many questions and loving the enthralling discussion of scientists. For Gallo, Jessen stood out. Decades later, he still remembers the amiable, blond young man whom he would see at all his talks. He remembers Jessen tenderly, calling him a "little angel." Jessen, to Gallo, was a man "you just had to like." So when Gallo got a call from him, desperate for new HIV drugs, he wanted to help. He put Jessen in touch with Franco Lori, a physician Gallo worked closely with, especially during the early development of AZT.

Lori, in turn, passed the question along to his colleague and close friend Julianna Lisziewicz. She had known Jessen during his brief stint of training at NIH. She was happy to help. Unfortunately, Lisziewicz didn't have much to offer. She had no magic drug. She didn't have data from a clinical trial. All she had to offer was an idea. Lisziewicz spoke to Jessen about a drug that she and Lori thought might have potential: hydroxyurea. It had been around for quite a while. It was, like the underpinnings of AZT, a cancer drug. Developed in Germany in 1896, hydroxyurea was approved by the FDA in 1967 for a specified group of cancers. The drug works by blocking a specific enzyme that's needed to form the single units of DNA, called deoxyribonucleotides. By blocking this one enzyme, the drug has been effective in battling cancer, psoriasis, and sickle-cell disease. The idea about HIV was that hydroxyurea would work similarly to AZT. By cutting off the ability to form new DNA, the drug was essentially cutting off HIV's technique of replicating itself.

Hydroxyurea also freezes the cell. Hydroxyurea keeps the cell from dividing, and HIV can't replicate in a cell that isn't dividing. With the virus frozen in place, other antiviral drugs can come in and attack. The beauty of using a drug that targets the cell instead of the virus is that the virus can't outsmart it. If given early in infection, the drug had the potential to keep the virus from getting a foothold in the body. At least that's what Lisziewicz and Lori hypothesized.

Even though the new drug hadn't been tested in an HIV patient, Jessen was desperate to try it. Lisziewicz had tested the drug only in cell culture, *in vitro*, in Robert Gallo's lab. They had yet to publish any results on hydroxyurea. There was no particular reason to think it would be effective in humans; many drugs that are promising in dishes warmed by an incubator fail when exposed to the delicate complexity of the human body.

When the chance of death is high enough, though, physicians will try anything. HIV was considered fatal. Under "compassionate use" exemptions, physicians were allowed to try other drugs, licensed for diseases other than HIV. Any drug could be repurposed to treat the virus. This approach favors the family-doctor style of medicine, often called cradle-to-grave medicine. Family physicians typically know their patients well. They are there for all their patients' firsts; they may know their patients' habits and quirks. They know if their patients can handle both the responsibilities and consequences of experimental drugs. Unfortunately, family doctors are in short supply. In an era when medical school tuition is soaring and interest rates for student loans are high, graduating physicians in the United States are drawn to higher-paid specialties, leaving a need for primary care doctors in nearly every state. In the early 1990s, 40 percent of medical school graduates chose to specialize in family medicine; today that number hovers around 8 percent.

A clinical trial involves a large team of researchers, statisticians, and administrators. Many people are surprised to learn that clinicians, those who directly treat patients, are often left out of clinical trial design. Instead, the initial spark for a clinical trial typically comes from a research lab where promising experiments are performed by PhDs, not MDs. These experiments are replicated, undergo peer review, are published, and then move into animal studies. The trial is designed to feed a statistical model. A clinical trial answers the big question: Can this drug work for the group that needs it? A family doctor answers the small question: Can this drug work for you? We need both systems and, perhaps, more cross-talk between the two. What works in a clinical trial will not always work for an individual and vice versa. Both fields inform each other, and ultimately, no matter how we get there, they both share the same goal.

Jessen, empowered by the international compassionate use provision, decided to give Andrew an experimental cancer drug never before tested in a patient with HIV. In doing so, he was taking a huge risk. He was risking his reputation as a doctor, his relationship with his patients, and, most of all, his relationship with the man he loved. The essential component of the doctor–patient relationship is trust. Patients, particularly those with life-threatening diseases, often put blind faith in their doctors, questioning little of their impending therapy. "You can convince a patient to agree to anything," says a candid physician at Massachusetts General Hospital before adding, "We have to impose our own ethical standards to make sure we don't ask too much of our patients." Informed consent by this measure often relies on the ethical standards of the physician. How much is explained to a patient, and how much they agree to, is defined only by the boundaries of a single doctor or researcher. It's a risky way to practice both medicine and science.

It was a risk that couldn't be taken lightly. Jessen decided he would leave Berlin to treat Andrew with the experimental regimen. He wanted seclusion. They found a house to rent on the beach in the North Frisian Islands off the northern coast of Germany. It was a seven-hour drive to Berlin; no one knew them there. Jessen feared that other doctors would find what he was doing absolutely crazy as well as unethical. Every day, Jessen rode a ferry to the mainland, flashed his ID, and picked up the experimental drug. On a strict timetable, he took Andrew to the mainland hospital himself, explaining that he was "sick" and needed a blood cell count. In their peculiar isolation, Jessen fought to maintain his composure. Andrew couldn't stand it and found every day of their two-month stay painful.

Andrew would occasionally ask, "What about you? Shouldn't you get tested?" Jessen was able to deflect the attention away from himself. It was a time, as he would later say, "when everyone died. . . . The only important thing was saving Andrew."

PART III

Treating the Berlin Patients

I wanted you to see what real courage is, instead of getting the idea that courage is a man with a gun in his hand. It's when you know you're licked before you begin but you begin anyway and you see it through no matter what.

You rarely win, but sometimes you do.

—Harper Lee, *To Kill a Mockingbird*

CHAPTER 13

The Second Diagnosis

Cancer. Timothy sat in a lonely, crumbling room at Charité hospital in 2006, trying to decide what to do. He had just learned that his cancer had come back. The hospital felt so old that it was hard to believe anything innovative could be happening within its walls. To enter a hospital and not know when you will leave, or if you will ever leave, is a horrible, hopeless feeling. Timothy, who had undergone three rounds of chemotherapy, hourly imagined the worst. Chemo made Timothy so ill that, each time, he didn't want to go through another round. The alternative, however, was worse.

He dreamed of Italy, where he had just taken what might be his last vacation. He had wandered Italy alone, traveling to Genoa and then along the coast for a few weeks. It was not an easy trip. After learning that his cancer had relapsed, his oncologist, Gero Hütter, had encouraged him to take a vacation. He had told him to get away and relax. The scenery was nice, but mortal feelings loomed. The time had come to decide whether to get a procedure that would be incredibly painful and life-threatening: a bone marrow transplant.

He knew how risky the procedure was. The doctor giving him a second opinion at a different hospital advised him against it, warning him of the risk. He had hoped the chemotherapy would work and he could return to a normal happy life. He had already endured so much. Why did he have to have cancer, too?

When he arrived in Milan, the rain was coming down softly on the cobblestone streets. He felt so alone. His boyfriend, Lucas, had stayed behind. Now that he was in the hospital again, the beauty of Italy came back to him. If he closed his eyes, he could still feel the warm welcome from his friends in Genoa, the taste of fresh seafood prepared just so. When he opened his eyes, the hospital glared at him. Cancer, not AIDS.

It started with the cold he couldn't shake. It felt as if he'd been sick for months, weariness, congestion, and pain flashing in and out of his life. HIV had become a minor concern. By 2006, HIV had become a manageable condition. The infection was no longer regarded as the death sentence it had been when Timothy was first diagnosed in the 1990s. He couldn't say the same about leukemia.

Just as when he was diagnosed with HIV, he heard the doctor speak to him slowly and softly in German. "Gibt es keine heilung." *There is no cure.* Timothy had been diagnosed with acute myeloid leukemia, or AML, a deadly cancer, with only about 25 percent of adults surviving five years after diagnosis.

As cancers go, AML is a particularly sneaky one. The cancer begins growing in the bone marrow. Hidden in the core of our bones lies a powerful soft, flexible tissue. The bone marrow is a font of the precious stem cells that grow to become all the mature cells we need in our blood. Every day, our bone marrow produces billions of blood cells. In addition to red cells, bone marrow also produces our white blood cells, or lymphocytes, which comprise our immune system. AML starts in the bone marrow, where the cancer spurs normal lymphocytes to grow wildly and replace the healthy blood cells. The cancer eats away at our immune system until we have nothing left; we can no longer protect ourselves. Or, from all appearances, you could say Timothy got a cold that he couldn't seem to shake.

Doctors don't know what causes AML. It might be exposure to certain chemicals, a blood disorder, or even having a weakened immune system. An HIV infection, which weakens the immune system, might be a factor. Being diagnosed with leukemia is a terrible blow for anyone. But for Timothy, the unknown effects of putting leukemia and HIV in the same body germinated a new kind of fear. He worried about the side effects of the cancer therapy as anyone would, but he also worried about having to stop his HIV therapy in order to be treated for his leukemia.

After his HIV diagnosis, he had told as many people as possible, to fight the isolation. Now he remained quiet. His many friends and the charismatic man he had once been, all faded into the distance that seemed a lifetime ago. All he had was the tiny dank hospital room, a boyfriend he loved to pieces, and Gero Hütter.

Following the very first meeting with Dr. Gero Hütter, in November 2006, he knew he could trust him. Hütter was a sharp contrast to Heiko Jessen. Sitting in Jessen's modern, white waiting room was loathsome to Timothy. Back in the mid-1990s,

Timothy would enter that patient room timidly. Jessen's patients loved him like a father and he loved them right back, hugging his way into their hearts. But for Timothy, who had never experienced warm, fatherly concern, this level of intimacy was unwelcome. He didn't want to be touched; he didn't want to feel Jessen's penetrating gaze, his soft concern, and his open heart. He wanted a physician, not a friend or a father.

Unlike Jessen, Hütter was focused solely on the clinical outcome. As a young oncologist, he had little experience in dealing with the emotional side of medicine. When Hütter met Timothy, he found him open, friendly, and very quick. He was tall and thin and looked nothing like a cancer patient. But this was clinically expected. Hütter had seen this phenomenon in many leukemia patients. They all looked healthy—until they started chemotherapy.

Their relationship unfolded with businesslike distance. Hütter explained very little of the science to Timothy at their appointments. He spoke instead of options, alternatives, and survival rates. Timothy responded well to Hütter's brisk nature, finding comfort in the impersonal yet scientific personality of the young oncologist.

Chemotherapy began. At Charité hospital, where Hütter saw Timothy, all leukemia patients under sixty years old are offered a stem cell transplant. Cancer patients often benefit from the transplant because the chemotherapy they take, while able to kill off cancerous cells, also kills off healthy cells. By transplanting new stem cells, the blood gets refreshed with a new supply of immune cells. The transplanted cells are not the embryonic stem cells that have been at the center of so much controversy.

Instead, hematopoietic stem cells are transplanted. These cells, although they can't form *any* cell of the body like their controversial counterparts, can develop into any cell of the immune system. The cells are found concentrated in the bone marrow or diluted in the bloodstream.

Cells can be donated from someone else, often a stranger the patient has never met, or concentrated and expanded from the patient's own body. For an allogeneic transplant, one in which the cells come from someone else, it takes time to match up the genetics from the person who needs the transplant to the person who's willing to have stem cells harvested from their bone marrow. The genetics that must be matched up are the HLA, the same set of genes that Bruce Walker was examining in his HIV research. Because we're matching up stem cells that form the immune system, the HLA, which are the genes that govern that immune system, have to be carefully checked to make sure there is a perfect match between donor and recipient. The search to find the right match for Timothy began, but of course he hoped he'd never have to have a stem cell transplant. He hoped the chemotherapy would work and he'd be back to his old self.

No one wants to get a bone marrow transplant. The procedure is dangerous. Before the transplant takes place, ablative treatment is given to prepare the patient's body. These drugs, like chemotherapy and radiation, make space in the bone marrow for the new transplant. Stem cells are then taken from the donor, either from their hip bones or concentrated from their blood. Cells can also be taken from umbilical cord blood after a baby is born, a rich source of hematopoietic stem cells.

Wherever and whomever they come from, the cells are delivered through a tube into the blood of the patient, and they make their way to the core of the bone, where they build a brand-new immune system. How well the transplant goes depends on how closely the donor and recipient are genetically matched. If the match isn't good, the cells from the donor will attack their new body and the result can be severe, even fatal.

Despite Hütter's conservative manner, he was a man not afraid to take considerable risk in his research. Hütter was planning a bone marrow transplant like none performed before. He had an idea, born from that paper he'd read a decade earlier. That paper, published in 1996, described the $\Delta32$ mutation and its protection from HIV infection. It was the catalyst for Hütter's experimental therapy but it was by no means a sure thing. Many a physician would not have wanted to take such a gamble. When an experienced virologist tried to explain to Hütter why it wouldn't work, he would only nod and acknowledge the risk; he still believed in the approach. Now he had to convince the hospital.

No HIV patient had ever received a bone marrow transplant in Charité hospital before. The hospital administrators said no, clinging to protocols from the 1980s, when AIDS was considered a death sentence. By this outdated logic, they reasoned that any patient with this deadly disease should be denied a costly bone marrow transplant, which would prolong life only for a limited time. Hütter, pressing his case, presented studies showing that HIV patients were now routinely given bone marrow transplants. He argued that HIV was no longer a valid reason to deny lifesaving treatment for cancer.

It was the first of many fights Hütter faced as Timothy's doctor at Charité hospital. Hütter was, after all, proposing a radical new method of treating HIV infection. Despite the fact that Timothy was Hütter's first HIV patient, the moment he met him, he had begun formulating a plan that encompassed not only treating his cancer but ultimately curing his HIV infection.

Like Jerome Horwitz before him, who envisioned a new way to cut off cancer from the cell, Hütter clearly recalled his first insight into the problem—that cold winter afternoon in his medical school library. HIV uses the cell in many ways, but when it comes to entering a human cell, HIV needs only two things: CD4 and CCR5. The target was therefore obvious.

CCR5, a gene that humans don't seem to need, is something that HIV absolutely needs. The plan was simple: Take out CCR5. The mechanism just as simple: a stem cell transplant. Timothy would receive chemotherapy and a stem cell transplant to fight his cancer. The opportunity was right in front of them. Instead of transplanting cells from any donor, they could find a donor with the Δ32 mutation. That way, when the stem cells formed a brand-new immune system, they would also form one that did not express CCR5 on the surface of T cells. These T cells, then, would be resistant to the virus. Even better, the virus would kill off the cells they could enter, selecting for a strong immune system capable of beating both cancer and HIV. It was a bold, elegant plan. Hütter fervently believed that it could work.

Charité hospital was, like many hospitals all over the world, a cutthroat, competitive environment for young doctors. They knew that permanent faculty positions were limited and only those who practiced both exceptional medicine and research would be promoted to one of the coveted posts.

Hütter felt the effects of this oppressive environment. As long as he possibly could, he withheld the details of his plan and of the presence of Timothy himself, for he knew that his competition would quickly try to squash such audacious plans from a junior doctor. Worse, for a long time, he hid Timothy's case from his own chief.

So it was no surprise to Hütter when infectious disease doctors at Charité hospital took up strong opposition. They argued that HIV can use receptors other than CCR5 to enter T cells. Consequently, transplanting Timothy with cells lacking the CCR5 receptor could not possibly keep HIV strains that use these other receptors from infecting Timothy. In fact, since Timothy had been HIV-positive for decades, it was more likely he harbored the CCR5-independent strains that become, for unknown reasons, more common later in infection.

The vast majority of HIV strains use the CCR5 receptor on the surface of immune cells to infect humans, but it is true that a small percentage of viruses use a different receptor: CXCR4. Viruses that use CXCR4 tend to be more pathogenic; they accelerate the virus's disease course in patients, causing rapid T cell death and general immune system destruction. Like the CCR5 receptor, the CXCR4 receptor influences how cells move around the body, but unlike CCR5, CXCR4 is an important receptor biologically. It is critical to how immune cells develop from the bone marrow and move into the peripheral blood. Humans who are born without the CCR5 receptor live normal, healthy lives.

However, we have no examples of a human able to live without CXCR4.

This is why infectious disease doctors were challenging Hütter's proposed therapy for Timothy. While eliminating CCR5 could undoubtedly contain some of the virus, the CXCR4 virus that Timothy likely harbored would still be able to grow. In fact, the approach could potentially result in a more dangerous HIV infection than Timothy had before receiving the transplant. And this was all supposing that it was even possible to get rid of CCR5 in patient cells, something that had never been done before.

Hütter, for his part, had far fewer studies to support his theory. There were no animal models he could point to where loss of CCR5 was linked to HIV protection. He worried that the reason no animal model had been published was because the experiments had failed.

His main argument rested on papers that had been published fifteen years previously. But the heart of his argument was not based on a model or theory; it was based on people, thousands of people who naturally lacked the CCR5 gene and yet lived healthy lives. It was based on the hundreds of people who, without the CCR5 gene, were resistant to HIV. Hütter's research did not address those people who, despite their lack of the CCR5 gene, still became infected with a CXCR4-using virus.

A stem cell transplant was different, he reasoned; they had the opportunity to reset the immune system, to turn back the viral evolutionary clock. He reasoned this because Timothy would be receiving not only a stem cell transplant but also a conditioning, or ablative, regimen to ensure that he didn't develop graft-versus-host disease. To avoid the disease, drugs are given to the patient before the transplant that dampen the immune system. In addition to suppressing the immune system, the conditioning therapy "makes room" in the bone marrow, by killing cells, for the new stem cells to expand. Between the chemotherapy and the conditioning regimen, Hütter believed it was likely that they were "resetting the immune system clock" by replacing so many cells, presenting a new opportunity to fight the virus.

He argued the case to Eckhard Thiel, the chief of transplantation medicine at the hospital. After months of hiding Timothy's case from his chief, afraid that both his idea and his patient would be stolen from him, he knew it was time to unveil his plan. As he sat in Thiel's first-floor office, he was nervous. He looked out the windows at the park just a few feet away. The paths were crowded with patients and their families outside on the mild day. As a junior faculty member, he had little sway, but what he lacked in seniority, he made up in passion. Although it would be expensive and have little chance of success, Thiel finally agreed. He was not sure Hütter's plan was reasonable, and certainly didn't believe that it could ever result in eradicating HIV, but he wanted to give it a chance.

Finding a transplantation donor would be a challenge. All potential donors for Timothy had to undergo additional screening to sequence their CCR5 gene. Only those with the mutant CCR5 gene, the Δ32 mutation, would be considered eligible. This severely whittled down the number of potential donors. This is the kind of experiment that would be difficult to do anywhere other than Germany. Unlike the United States, Germany maintains a large database of bone marrow donors. In 1991,

German donor registries received funding to build an extensive database of donors. In that single year, the number of donors went from around 2,000 to more than 50,000. Today the German registry, ZKRD, is the largest in the world and has access to more than 19.5 million patients worldwide. Through the registry, 75 percent of all patients receive a matched donor within three months; overall, 90 percent of patients are matched to a donor. Compare this to the United States, where only 65 percent of patients will ever find a donor through our national bone marrow registry.

Another advantage Germany has in the hunt for the mutant gene in the haystack is the high rate of this particular mutant gene in Europe. Fourteen percent of Europeans carry one copy of this mutation in their genes, a freakishly high number when compared to the rest of the world. In about 1 percent of all Europeans, both copies of the CCR5 gene are mutant. When both copies are mutated, the body can't make the CCR5 protein, and HIV is left standing at the door, unable to turn the key.

Timothy was born with only one functional copy of the gene. That one copy was producing all the CCR5 that HIV needed to enter. If Hütter wanted to shut HIV out, he had to stop Timothy's intact CCR5 gene. To do that, he had to find a donor who had no functional copies of the gene. That meant only 1 percent of the population was eligible.

No one knows why Europeans have such a high propensity for mutant CCR5. The mangling of CCR5 arose as a single mutating event about 700 years ago. This is considered a young mutation. Compare it to one of humanity's eldest mutations that allowed us to properly convert plant fatty acids some 85,000

years ago. Some believe bubonic plague was the original instiga-
tor for the mutant CCR5. Plague, the "Black Death," the pandemic that killed an estimated 100 million people in the Middle
Ages, is caused by a bacterium, *Yersinia pestis*. These bacteria
hijack macrophages, an immune cell that expresses CCR5. Some
researchers think the bacteria enter our cells by a means similar
to HIV, through our CCR5 receptor. However, studies on mouse
susceptibility to the bacteria, in animals both with and without
the CCR5 receptor, disagree on this point. The bubonic plague
wiped out one-third of the European population. Given this
widespread death, it makes sense that if a tiny mutation is able
to keep the bacteria at bay, the mutation could become influential. Other theories have focused on smallpox, a virus whose
entry into our cells remains mysterious. Some evidence exists
that the poxvirus uses CCR5 to enter human cells. Whether
powered by the bubonic plague, smallpox, or some driving force
not yet identified, those with the Δ32 mutation had a survival
advantage so powerful that their children carried that mutant in
their genomes, to be passed down the generations. That survival
advantage would lie dormant for centuries until the AIDS epidemic woke it from its slumber, extending its ancestral protection across time.

As the Charité team scoured Western Europe in search of
Timothy's perfect HIV-resistant match, Timothy felt lucky to
have Lucas in his life. He didn't want to have the frightening
stem cell transplant at all and especially didn't want to go through
it alone.

CHAPTER 14

The Compassionate Use Exemption

C hristian sat near Timothy in Jessen's crowded waiting room. The two men, the Berlin patients, were scheduled to see Jessen that afternoon. Each was unaware of the other's significance to medical history. It was 1996, long before Timothy's cancer diagnosis. They had been diagnosed with HIV only a short while ago. They were close in age, had a similar build, and shared personality traits: reserved, a touch sensitive. They didn't even know each other's names. They sat in the waiting room, like many other patients, studiously avoiding eye contact.

It had been three years since Jessen had treated his boyfriend, Andrew, with hydroxyurea. Yet, the drug remained a mystery; there was only anecdotal evidence that the cancer drug was an effective therapy for HIV. Jessen's limited experience with the drug was positive. Andrew had, thus far, survived. Returning from their sojourn to the northern islands of Germany, his viral load had lowered, and his T cells increased: encouraging evidence that the "intervention" had been a success. However, AZT was able to lower virus and increase T cells in the short term, before the virus mutated around it. The real test for Andrew's intervention would be its ability to keep him alive in the long term. Andrew and Jessen's relationship, however, would not last long-term. Returning to Berlin, Andrew had broken up with Jessen, stopped the radical therapy prescribed by him, and left Germany for Spain. He bounces around the world today still healthy and frequently dating doctors. We'll never know what effect the novel drug has had on his survival. Jessen was, and in some ways still is, heartbroken.

Nevertheless, Jessen believed in the hydroxyurea, openly prescribing it to a select group of patients. It wasn't a large clinical trial. Instead, it was a small trial, the kind that still occasionally happens in family doctors' offices. Part of the reason the trial was small was because patients had to be carefully selected for it. They had to be recently infected and very responsible. It was a drug that had to be taken at specified times every day. Jessen also had to do significant follow-up work on patients and needed to be sure the patients would show up to appointments. It was advantageous for him to have a relationship with anyone taking the drug. Jessen was interested in how the drug could be combined with the "hit hard, hit early" strategy touted by David Ho and others, to beat down the virus, eradicating it from the body. There were still no large-scale clinical trials of hydroxyurea, but the strategy intuitively made sense to him. He would use the powerful drug to control the virus at the earliest opportunity, before it had a chance to entrench itself in the body.

When Jessen prescribed Christian hydroxyurea, he did it carefully. Christian had been diagnosed very early in infection. He was a responsible student. He had been coming to Jessen for a year as his family doctor. Jessen felt he could trust him. Christian, for his part, didn't think anything of the experimental drug. He didn't think to question it. He wasn't particularly interested in the science behind it. All he knew was that he had a potentially fatal disease and he needed to take medicine. He dreamed of being "the exception to the rule." That somehow this experimental combination would work and he could actually be the first person cured.

When Timothy first saw Jessen, he had harbored HIV far longer than Christian. He had been infected nearly a year earlier and hadn't known it. Because the early symptoms are similar to the flu, this happens frequently; a recent study found that 44 percent of HIV-positive gay men in the United States don't know they're infected. Jessen didn't think to prescribe hydroxyurea to Timothy. After all, what were the chances that you could clear out the virus when it already had a foothold in the body? Timothy got a very different regimen of antiviral drugs and went on his way. He wasn't unhappy with the medicine, but he was displeased with Jessen's warm and fuzzy personality.

Jessen's hydroxyurea trial, given outside the typical context of a structured clinical trial administered by a hospital, was radical. Once HIV was no longer considered a death sentence, HIV clinical trials changed. Similar to how, if you're dying of thirst, you'll drink almost anything, even your own urine, early HIV trials were desperate affairs. Today, HIV patients in the States and in Western Europe are no longer dying of thirst. They can afford to be picky about what treatments they take and what clinical trials they participate in. Nowadays, HIV clinical trials typically involve busy infectious disease specialists who have to keep all their treatment participants on identical structured regimens to maximize the statistical power of the clinical trial. Jessen pursues a very different approach to medicine, one that puts the focus on the patient rather than the medicine.

As Christian took the hydroxyurea, he imagined a toilet bowl commercial that was shown on TV when he was a kid in the 1970s. In the commercial, a large bowl of blue water is shown in the foreground. A tablet of toilet bowl cleaner, like a giant pill, is dropped into the blue bowl. Like magic, the blue water becomes sparklingly clear.

As he took his hydroxyurea every morning, he visualized the commercial. He imagined the drug acting like the toilet bowl cleaner. He dropped it in his body and, like magic, it cleaned him of the virus. The visualizations were a comfort to him. He believed not just in the science, perhaps not even mostly in the science. He believed in the spiritual act of taking his medicine—letting both the drug and his good thoughts scrub his body clean of HIV.

CHAPTER 15

Three Deadly Diseases Move In

The first weeks after he was diagnosed, Christian had taken his medication religiously, never missing a dose. To ensure this, he wrote out an elaborate schedule, plotting out meals and medications in excruciating detail. One of the most difficult parts of the schedule was not being able to eat before taking the medication. Christian had been prescribed didanosine, or DDI. This early HIV drug came as a large white tablet, exceedingly fragile to the touch. The tablet needed to be crushed in water before being taken. It had a terrible taste to Christian, both sweet and acidic, rank with the artificial flavor of mandarin oranges. Any drug taken by mouth is absorbed by the body at a lower rate than, say, directly injected into a vein. The problem with DDI is that while it's absorbed quickly, a low percentage of the drug actually gets where it's needed. This is called bioavailability, and for DDI the rate is particularly low; only 42 percent of it is properly taken up by the body. Compare this to drugs given intravenously where the bioavailability is 100 percent. For DDI, taking the drug with food reduces the bioavailability by another 25 percent. Therefore, to get the highest drug concentration possible, it has to be taken on an empty stomach.

This unfortunately meant that Christian had to miss breakfast, waiting hungrily until his morning break at 10:00 A.M. He sat in the cafeteria, surrounded by other history students. Before his diagnosis, he would normally only drink coffee at this break. Now, he was ravenous, eating a full meal. Such a dramatic change in his eating habits brought on questions and teasing from the other students. Yet Christian remained mysteriously silent. He couldn't possibly tell them the reason for the change. Instead, he tried to shrug it off, feeling exposed and embarrassed.

Worse than the grueling schedule was how the drugs made him feel continually tired and nauseated. The sickness was almost unbearable, but worse was the feeling that he was different from everyone else. It was an isolating feeling. Especially since Christian had few close friends in Berlin he could speak to. He knew no one else who was HIV-positive.

The drug regimen was interrupted for the first time in August. He had been taking his new anti-HIV drugs for only a little over two weeks when he developed epididymitis, a painful swelling of the tube that runs along the back of the testicles. Desperately in pain, he rushed to the hospital. He forgot to bring his anti-HIV medicine. As he was admitted to the hospital, he explained his situation to the physicians, hopeful that they would be able to provide the drugs he forgot at home.

The physicians were confused. They couldn't understand why a person so early in HIV infection would be taking antiviral drugs. They had never heard of the "hit hard, hit early" strategy and knew nothing of Jessen's plans to sneak-attack the virus in Christian. Even more puzzling, they couldn't understand why Christian would be taking a chemotherapy drug, which is what hydroxyurea is, to treat HIV. They explained all this to Christian, telling him that his family doctor was likely not a very good one.

Christian spent seven days in the hospital without his medicine. For a whole week, he didn't have to consult his complicated handwritten chart of when to take each drug and when he could eat. It was a relief to be free of the drugs, but it was also harrowing. He worried that without the medicine he could die. Released from the hospital, he quickly went back to his routine. He found strange comfort in taking the pills that made him sick. He practiced his visualizations.

That year, Christian was diagnosed with HIV, epididymitis, and hepatitis A. For Christian, who had never experienced illness before, it was a year of hospitals. He couldn't get into a rhythm. It seemed that as soon as he started to get used to taking his HIV medications, he would get diagnosed with something else, be admitted to the hospital, and have to stop taking the drugs. As he lay in a hospital bed in Berlin with hepatitis A, he felt overwhelmed. Then he learned that his grandmother had died.

With his grandmother, he had always felt he was special, the favorite of his cousins. As a small child he adored her. She was strong and kind. As he grew older their bond deepened. In many ways, he felt that she understood him on a level he couldn't quite define. Although he had never told her he was gay, he's sure she must have known. His sister broke the news of her death to him gently, but he felt as if his world was crashing down around him.

He wrapped his arms around himself and dropped his head. He let out one soft sob and then was quiet. Just as when he was first diagnosed with HIV, he couldn't speak. He had no words. He couldn't attend his grandmother's funeral. He would have to stay in the hospital. His body ravaged by viruses and bacteria, his beloved grandmother dead. He tried to tell himself that things would get better, but in his heart he felt no relief, no comfort. It was the worst year of his life.

Meanwhile, inside Christian's body the virus was rising and falling. His body had begun fighting back. In a few weeks, he would get the best of his hepatitis and be able to leave the hospital.

It rarely snows in Berlin. It rains, it sleets, it sprinkles, but rarely does the city experience an honest-to-goodness blizzard. However, in November 1996, the city was buried under several feet of snow in one of the largest storms ever to hit the city. Everywhere, traffic came to a halt, schools and businesses closed early, and children, delighted by the promise of an extended holiday, played in the streets.

Christian stood by the window in his student dormitory in the former East Berlin. He was a student at the Free University of Berlin, studying history. His dorm room was small, the twin bed crowded against a desk, but it was completely his own. He even had his own bathroom. The room smelled of new paint, having just been renovated. He was the first person to live in it since the squatters had left after reunification. For nearly a decade, every sort of people occupied the apartment, all living rent-free. Now, years later, the floors had been stripped and the walls painted. The room matched how he felt inside: clean, new, purged of virus.

The room had one window, awkwardly framed. It was a small pane that came down only to Christian's chest. He looked out into the evening sky, watching the snowflakes circle down past his window and land in the yard below. The sky was dark gray, the evening creeping into daylight hours as the calendar approached the longest day of the year. Christian had been sick for six months. He had endured endless mornings of retching and dry heaves. He had suffered extreme exhaustion, could barely work, and had kept a chilling secret from friends and coworkers. Now, for the first time in months, he was beginning to feel like himself again.

He leaned against the wall and picked up the three pill bottles balanced on the narrow window frame. The labels were sticky from the condensation that had formed on the inside glass of the window. He twisted the bottles slowly back and forth in the palm of his hand; it was time to take his medication.

After the first few terrifying months, he began to become more relaxed. He sometimes missed pills, always rationalizing to himself that there was a reason, that it was human to make mistakes. "I have to make this meeting," he would tell himself. "I don't have time to go back to the house."

But after he missed his pills, he became nervous. He wanted to be a success. He wanted to be, in his words, "the exception to the rule." He wanted to be cured. He didn't want his family and friends to suffer simply because he couldn't seem to swallow his medication. The remorse rose up in him each day and it seemed the only way to push it down was to swallow the drugs he held in his hands. But today felt different. He looked out the window at the buildings covered in white, the quiet street below him. It felt like a new world out there, his city freshly bathed in freezing water.

Inside, he felt as clean as the pure, white snow outside the window. It was almost as if he could feel his own health rising up and filling his body. He had suffered some of the most difficult months of his life. Pain, both physically and emotionally, had consumed him. But now, standing at the window, he felt what he would describe as a "moment of clarity." He wasn't quite ready to let the drugs go completely; that would come a month later, at Christmas. But this was a defining moment. A spiritual moment.

He looked out the window, lightly touching his forehead to the cold glass. Many patients in his position would have been preparing to die, but not Christian. He opened the window and let the freezing cold air rush in. Despite all evidence to the contrary, despite the fact that the virus had rebounded in his body when he had stopped taking his medication before, in his heart, he knew that this time the virus wouldn't return. He let the pill bottles fall to the floor as he wrapped his arms around his body, hugging himself tightly. His body tingled with feelings of unreasonable hope. *I'm healed,* he thought.

CHAPTER 16

The Comfort of Family and Strangers

Timothy was a teenager. His best friend was Samantha, a girl he'd grown up with and cared about fiercely. He knew it was time to tell his mom he was gay. Samantha kept reminding him that it had to be done. Yet he couldn't seem to muster the courage. Every time he decided it had to be done, he would lose his nerve. Timothy was never one for conflict; it simply wasn't his personality. His mother's Christian faith made the situation more delicate. He knew she wouldn't be pleased. He also knew she needed to know. He loved his mom; he couldn't bear her not knowing such a fundamental truth about her son. So he wrote her a letter. Some subjects are better described in words on a page. Some people are more apt to listen when they're alone with a letter.

The letter was difficult for Timothy's mom to read. Although she couldn't know it at the time, it would be only the first in a line of letters that Timothy would send her when he had bad news. It was always easier for him to share news by writing it rather than face-to-face. Even though he knew it was better to tell certain news in person, he dreaded the feeling of disappointing his mother. When Timothy's mom was finished with the letter, she called her own mother, Timothy's grandmother, back in Idaho. She had to share the news with someone. A similar pattern would take shape when Timothy decided to tell her he was HIV-positive a decade later.

Timothy, like Christian, was close to his grandmother. He could never acknowledge his sexuality directly to her, much like Christian, and yet, also like Christian, she already knew. There was an unspoken understanding at the foundation of their relationship. Both their grandmothers seemed to know them better than they knew themselves.

Christian was nervous as he prepared to tell his parents, face-to-face. He was eighteen and had known forever. Christian's family was very different from Timothy's. He had a close relationship with both his parents and felt their love and support. He knew they might have some concern, but he also knew they would accept him. True to form, Christian's parents embraced him seconds after hearing the news. They didn't worry or make him feel abnormal.

Their response a decade later, however, when Christian told them he was HIV-positive, was not as calm. Christian was nervous about telling them. He had known for a month and had told only a very few friends. His first thought after being diagnosed had been his parents. He was so anxious to know what they would think of him; how their opinion might be changed; if they would look at him differently. On his three-day visit home, he waited until the last possible moment, finally stirring up his courage. His mom burst into tears at the news, and his dad supported her weak frame. Christian couldn't stay to watch; he left quickly after telling them, retreating back to his apartment in Berlin. His relationship with his family suffered under the strain. His parents were worried for their son's life, for they saw HIV as a death sentence. They had other worries, too: They lived in a small town and couldn't help but wonder what neighbors would think, how it would be perceived in their community. But they loved their child more than any of these fears. Their relationship eventually healed.

Jessen's family was much like Christian's. They were both close-knit families from small towns. Jessen's parents knew early on that he was gay, so when Jessen came out to them, they were unsurprised. As it happened, all of their children were gay, two sons and a daughter. All of their children would leave the farm on which they were raised, drawn to the city of Berlin. Incredibly, all of their children would find work shaped by HIV.

Supportive families like Jessen's and Christian's make a big difference in the emotional health of a gay person who's come out and also influence their physical health. Studies performed in multiple ethnic backgrounds all show the same result: If an individual's family is unsupportive of their sexual orientation, the person is more likely to binge drink, to use illicit drugs, and to suffer from depression. In contrast, gay men with supportive parents are three times more likely to use condoms during sex and be regularly tested for HIV. At its heart, this intuitively makes sense. Our parents help shape the way we see ourselves as adults. Therefore, if our parents don't support us when we reveal our sexuality, our self-esteem takes a blow. We have less reason to take care of ourselves. This may be why the suicide rate is so much higher, 8.4 times higher, in gay and bisexual adolescents than in their heterosexual peers. Coming out is a precarious time in the life of a teenager. We can only hope that the explosion of new studies looking at the profound influence that parental approval has on the health of gay and bisexual children will influence the behavior of future generations.

After Timothy was diagnosed with HIV, his behavior took a sudden turn. He found the dating world split between those with the virus and those without. It was almost as if there were two populations of gay men—those condemned to die and those free to live. He belonged to the first group. The dating practices of that group were different. They could pick each other out at a crowded bar by the lesions caused by Kaposi's sarcoma and by sunken cheeks, symptomatic of the wasting disease caused by AIDS. Even if a person with the virus eats normally, and many cannot, diarrhea, vomiting, weakness, all contribute to a wasting away of muscle. This is called cachexia and is not unique to AIDS. Those with end-stage cancer often suffer from cachexia, becoming weaker and weaker, no matter how much they eat.

For HIV, the pattern is complex. Those with the virus often experience fat redistribution. Lipodystrophy (where *lipo* stands for fat, and *dystrophy* an abnormal change) is a distribution of fat, often occurring in the fat of the face, resulting in sunken cheeks. Puzzlingly, we don't know exactly how this occurs. It seems to happen to a larger extent to those on antiviral therapy. Our current thinking is that therapy taken to beat back the virus may also be damaging cellular mitochondria. Mitochondria are the "power packs" of the cells, tiny organelles inside each of our cells that pump out the vast majority of energy our cells need to function. The therapy taken to target the virus also appears to target the mitochondria in fat cells, particularly those in the face. Without the mitochondria, these cells die off, giving the cheeks a hollow, sunken look. HIV-positive people with this condition, while it is not dangerous, find themselves stigmatized; after all, the mark of the virus is out where everyone can see it, on their faces. Today, we've been able to find new drug combinations that are less likely to create lipodystrophy. This isn't always effective, since sunken cheeks are not exclusively tied to drug therapy. Others have turned to implants and fillers that can hide sunken cheeks. There are even patient-assistance programs available to provide these fillers so that people with little money don't have to live with HIV's telltale signs.

But in 1996, after Timothy learned he had HIV, none of these advances existed. It was easy to recognize those with HIV. When dating, Timothy found himself looking for sunken cheeks. He purposely sought out men with the red and purple bruised marks of Kaposi's sarcoma. For Timothy, he looked for these marks as the signposts of his new community. He wanted to be responsible. He didn't want to infect anyone else with the virus, so when he went to bars, he sought those who also had HIV. It was a new kind of segregation for Timothy. It made him feel as if he had lost part of his identity. He was no longer the gregarious partygoer he had once been. Instead, he sulked in bars, looking for people he could identify with.

One night he went to a bar in Berlin down the block from Jessen's clinic. At street level, the bar seems normal. However, in the back, there's a crawl space that leads down to a second bar that never sees daylight. It's called a dark bar because it lets in no light, either natural or artificial. Timothy picked this bar one particular evening to bask in anonymity. It was a place where he didn't have to worry about how he looked, or how anyone else looked, only how he felt. It wasn't the kind of place you went to find your soul mate. It was all about sex at the dark bar. In the early morning hours, he touched the face of a young man sitting across from him. He knew what he had come there for, but still, he wanted to talk. So he did. He sat in the dark bar, talking to a man he didn't know. His feelings seemed to pour out of him, a rare experience for Timothy. The comfort of being invisible was a powerful drug, making him talk about things he rarely admitted to himself. He left that night with a man who would not be simply a one-night stand but instead his closest friend, his soul mate, the love of his life. He couldn't know that the man he held in his arms that night would one day, years from then, stand beside him during the most traumatic time of his life. Lucas would become everything to him, an essential part of his treatment and eventual release from HIV. But all of that was years away.

CHAPTER 17

Timing

I t was almost Christmas. Christian made his way home to the small town he grew up in, in the German countryside. He was excited to go home and see his family. Christmas was always special in his house. He was close to his parents and sister and missed them, living hours away in Berlin. Christian's mom greeted her son with a tender embrace. She worried for him, her only son, who harbored a deadly disease.

Arriving at home, Christian found the house was decorated in the familiar way he remembered from his childhood. His mother had begun baking weeks before Christmas, filling the house with cookies and sweets. Garlands twisted their way around the house, and everywhere lights and candles softly glowed. On Christmas Eve, they decorated the big spruce tree with lights and ornaments, filling the room with the scent of evergreen. They ate sausages and potato salad on Christmas Eve, a simple menu compared to the feast coming the next day. On Christmas Day, they ate a large turkey, the extended family all coming to Christian's parents' house, which was filled with delicacies. All year,

Christian craved the green salad they would eat on Christmas Day. Christian's uncle grew the endive himself, bringing the salad to the feast fresh from his farm. "Of course you can buy the salad, but it isn't as special," says Christian, remembering how precious the tradition felt that particular year.

As he sat at home that vacation, he felt different, more confident. On the snowy day in November, he had felt clean, pure, purged of the virus, and yet he hadn't had the courage to completely stop taking the drugs. Off and on he had taken them, unsure of what he should do. He had refilled his antiviral drugs before coming home for the holiday. Now here they sat in his childhood home, seemingly challenging his inner self. If he really believed he was free of virus, he reasoned, then he could stop altogether, be free of the drugs.

A shy man who usually did exactly as he was told by his doctor, he had developed a new confidence in himself. He was ready to go against Jessen's prescriptions even though he trusted his doctor completely.

The final moment of his decision to stop taking the medication came just a few days after Christmas. Christian stood in the doorway of his childhood room in his parents' home as his mother stopped by, nonchalantly asking him about taking his medication and how it was going. Leaning against the wooden door frame, Christian told her simply, "I can't continue." He said no more; he couldn't quite bring himself to give her the full details of his own powerful feelings of recovery. Although worried, she trusted her son to make his own decisions. He had spoken only a single sentence, but somehow saying it out loud made it feel real. Christian felt committed by his statement to his mother. It empowered him to make the final decision for himself. From that day on, he no longer worried about schedules and when he needed to take his medication. He was free. He would never again take antiviral drugs. As Christian contemplated his new freedom, Jessen stood miles away in his clinic in Berlin. The clinic stayed open on Christmas Day and Jessen was working.

At the end of 1996, Jessen nervously called Julianna Lisziewicz. He had known Lisziewicz for only a few years, but they had become close friends. He had met her while training in Rob Gallo's lab at the NIH. Lisziewicz loved talking to Jessen about Germany, where she had gotten her PhD at the Max Planck Institute in Goettingen before joining Gallo's lab for her postdoc. In the 1990s, Gallo's lab was where every young scientist interested in viruses wanted to be. It was "likely the largest laboratory in the world," says Lisziewicz, remembering the fast-paced energy.

Lisziewicz, far ahead of her time, was interested in a gene therapy approach to treating HIV. She devised a clever idea based on basic cell biology. Genes are expressed in cells by tiny single-stranded pieces of RNA called messenger RNA, or mRNA. It's called a messenger because it carries the blueprints of the cell copied from the DNA of the cell's nucleus to the construction plant, where that message directs the proteins the cells need. Lisziewicz's idea was to make small pieces of DNA complementary to the mRNA that HIV uses to direct its genes. Those little bits of DNA could bind up the viral RNA, thus stopping the ability of the virus to make more of itself. These small bits of gene-stopping DNA are called antisense oligonucleotides. Liszie-wicz's results looked beautiful in cell culture, and the project was quickly moved to clinical trials.

In six months, Lisziewicz had gone from quietly doing a side project that she knew Gallo wouldn't approve of, to being ap-pointed head of the antiviral unit. In that position, her contacts and influence grew. Lisziewicz could both conduct successful research and raise money. In 1994, it seemed that the right time to leave the NIH had arrived, and she started a nonprofit com-pany, the Research Institute for Genetic and Human Therapy (RIGHT), with Franco Lori, a colleague from Gallo's lab.

All through these changes, Lisziewicz stayed close friends with Jessen. In 1993, when he was desperate for some kind of experimental therapy to give his boyfriend, Andrew, she was there, providing a drug that no one else had yet used for HIV. When Jessen decided he wanted to perform a small trial of the drug at his clinic, Lisziewicz was excited. Jessen was in a perfect position to conduct such a trial. His clinic was, and still is, unique because of the large number of patients who are diag-nosed soon after they are infected with HIV. Jessen's patients, because they trust him, will come to him early with their fears and possible risk. For a clinical trial, the setup is near perfect. In fact, Lisziewicz says that at the time, "no one else in the world could have done this trial."

Not only did Jessen have the right patients, he was intensely interested in early intervention for HIV, inspired by his own personal experience with Andrew. In only a few months, Jessen had enrolled 13–14 patients handpicked to ensure that each patient was started on the experimental drug, hydroxyurea, soon after they became infected and was responsible enough to adhere to the drug regimen. With such a small number of patients, each one was essential to the trial. That's why Jessen was nervous as he called Lisziewicz at the end of December. One patient had gone awry: Christian.

Jessen related what Christian had told him. Over Christmas, he had decided he would no longer take the medication. Coming back to Berlin after Christmas, he laid the still-full pill bottles on Jessen's desk, turning in his weapon against HIV. Lisziewicz was upset. "He can't!" she cried. There was no way the virus could be eradicated from his body after only six months. Additionally, the six months had been fragmented; he'd already had to stop taking the drugs a few times, resulting in the virus rebounding throughout his body. Jessen agreed with everything Lisziewicz said, and he told her he had tried to convince Christian already, but it was no use.

Jessen was never one to pressure his patients. He believed that the most he could do was give his opinion; it was up to the patient to decide what he wanted after that. While speaking with Christian, he was surprised at his conviction. Christian was shy and mild-mannered. He had followed Jessen's instructions closely at the beginning. To see him so sure of his decision to end therapy made Jessen consider his own position. He could ask Christian to restart therapy, even explain the danger if he didn't, but he wouldn't pressure him into it. Instead, he would calmly ask him to keep coming back. He explained that it was important to keep testing for the virus so they could restart therapy when the virus began coming back. Christian agreed but intuitively knew that the virus was gone. He wasn't worried about Jessen's warnings; after all, the drugs had already worked, at least in his mind, and he was glad to leave the clinic with only minimal chiding.

Lisziewicz respected Jessen; he was an excellent doctor. She knew Jessen would always be a doctor first and scientist second. So she expressed her disappointment and they moved on, making new plans for the rest of their patients.

CHAPTER 18

Transplanting

Deciding to end therapy as Christian did was the kind of decision that Timothy could never have made in 1996. That year, he switched from taking AZT to a new drug: retrovir. It made him feel just as awful, but he was pleased to be taking something new. Pleased, even excited, but he had absolutely no sense that the virus was gone from his body. Eventually, in 2007, Timothy would have his own Christmas moment, as influential as Christian's. But it wouldn't happen in Berlin; it would happen in Idaho. And it would not be so cheerful.

Timothy loved living in Berlin, but he missed his family. Every year he tried to travel back home during the holidays. In 2007, he was particularly anxious to get home. His grandmother, whom he was close to, had pneumonia. He was worried about her. The flight home gave him long hours to reflect on his leukemia—diagnosed earlier that same year. Being HIV-positive had become such a constant in his life, shaping his identity, his relationships, and even his work. Now this . . . At first it had been difficult to accept that he had cancer, even more trying to contemplate the treatment and the risk. But now this Christmastime, he was in remission. He had beat back the cancer. He was thankful that the bone marrow transplant, a risky procedure, had gone well.

Unlike Christian, who exuded health and felt well-being at his very core, Timothy felt exhausted and sick. It was hard to remember a time when he hadn't felt that way. Had it been two years? Three?

He'd had a resurgence of spirit following his bone marrow transplant for the leukemia. He'd gone back to work, started working out at the gym again. He even found himself playfully flirting with cute straight men just as he used to in the old days. He was beginning to feel like his old self again. As the months passed, that healthy feeling eroded. Recently, he'd been suffering from other, random illnesses. He'd had shigellosis and a norovirus infection before leaving for Idaho.

In addition, his relationship was crumbling. Lucas wanted to see other men, indeed had begun to do so. It was a hard blow for Timothy, who loved his boyfriend, even more fiercely after watching him stand by his side through the leukemia treatment. Timothy couldn't just walk away. They left it as it was, a relationship clinging to the edge of a cliff, about to let go.

Timothy's childhood was filled with memories of traveling with his mother from Washington State to Idaho at Christmas. As an adult, he kept up the happy tradition, trying to make his way home even when he lived thousands of miles away. This Christmas, he had brought Lucas's niece, Sophie, who was excited to visit the United States. Even though Lucas wasn't there with him, he found solace in having a part of Lucas's family with him.

The family gathering was big in Idaho, the house packed with aunts, uncles, and cousins. Timothy was happy to be home and trying to ignore his deteriorating health. He didn't want to ruin the Christmas holiday with his growing list of medical conditions. *I must have pneumonia again*, he thought, depressed at his continued sickness. Timothy's mother, who was worried about her ailing son, took him to the doctor. The nurse drew his blood, and the physician told Timothy exactly what he expected: He had pneumonia. The diagnosis was a relief; with his cancer so recently treated, he knew the results could be far worse.

That fear would become crystallized the next day, when the results of his blood work came in. The news crushed him. The cancer was back. Timothy knew what this meant. He'd have to undergo either another round of chemotherapy or possibly a second bone marrow transplant. The idea of reliving the excruciating procedure was traumatic in itself. He quietly told only his grandmother, the strong matriarch, knowing that the news would find a way to trickle down to the rest of the family. *It's starting again*, he thought to himself. Christmas came and went quietly.

Inside Timothy's body, a few cancer cells had made their way past the irradiation and bone marrow transplant he'd had less than a year before. Now those cancer cells were growing uncontrollably yet again. When chemotherapy doesn't wipe out nearly every cancer cell, the disease comes back. The cancer Timothy had, acute myeloid leukemia, or AML, grows abnormal leukemic cells quickly in the bone marrow. The overabundance of white blood cells created by the cancer means there is now a profusion of immature white blood cells called blasts. A bone marrow transplant to treat AML is usually successful at making the cancer go into remission. Timothy was scared. The fact that the cancer had reappeared was a very bad sign.

He returned to Berlin sad and lonely. And now he could feel a shift in the attitude of Hütter and the other doctors. While the blood test that Timothy had in Idaho was a worrisome sign that the leukemia had returned, as indicated by the abnormal number of blasts, the official diagnosis wouldn't occur until he received a bone marrow aspirate. To do this, a needle is put inside the bone, and a small amount of fluid and cells are removed from the core: the bone marrow.

When Hütter saw the bone marrow aspirate, he could tell that it was full of blasts, a positive diagnosis that the leukemia had returned. Even worse, the leukemia had become more aggressive. Now the blasts were found not only in Timothy's bone marrow but also in his lymph nodes. With the first transplant, there had been a spark of excitement. Hütter had exuded optimism and hope. He had believed wholeheartedly that they would cure Timothy. But now, with Timothy's relapse, the mood changed. Before the first transplant, Timothy had a 25 percent chance of surviving 5 years. Now, with the advanced state of his cancer, his survival rate had dropped to less than 11 percent. As he sat in a small café in Berlin, Hütter's face became grim when he remembered that terrible time. It was "a sentence to death," he explained.

It was a painful reversal of fortune from Hütter's perspective as well. After the first stem cell transplant, he was elated to find Timothy bouncing back so quickly, resuming his normal life. Even more impressive had been the effect of transplanting from a donor naturally resistant to HIV. They found that only two months after the transplant, every discernible cell in Timothy's body expressed the Δ32 mutation, the mutation responsible for keeping HIV at bay. Also incredibly, the virus in Timothy's body had disappeared following his stem cell transplant, and importantly, it hadn't returned. This was especially impressive since Timothy's body had contained very high levels of virus only days before the transplant: He had a billion viruses in each milliliter of his blood. A milliliter is a very small volume of blood; it takes about five of them to fill a teaspoon. Most HIV patients don't have such extreme levels of virus in their blood because they take antiviral drugs that keep the virus in check. Timothy, however, had to stop taking his medication after he was diagnosed with leukemia due to toxicity complications with his chemotherapy. Once he stopped the medication, the virus grew uncontrollably, replicating to extreme levels. Following the stem cell transplant, the virus became undetectable. From a high of a billion copies per milliliter, the virus had dropped to nothing.

What made Timothy's vanishing virus even more impressive was that he hadn't restarted his antiviral therapy. His immune system was keeping the virus in check with no outside help. His T cells, too, while still peaking and dipping, seemed to be on their way up. Hütter knew it was too early for celebration; there were many problems that could come up. But he couldn't keep himself from being excited. These key clinical parameters were remarkable. Unheard of.

Timothy already knew from previous experience that, like the vast majority of others infected with HIV, if he stopped taking his medication, the virus in his body would rise up and his T cells would be killed off. The fact that a stem cell transplant from someone naturally resistant to HIV could turn this trend around was revolutionary, even if the therapy ultimately didn't cure. Hütter proved to his colleagues at Charité hospital that it was possible to change the genotype of a patient to resemble one naturally resistant to HIV. This strategy plummeted the virus in the body to undetectable levels and was beginning to elevate T cells. What he couldn't prove was that a more aggressive strain of HIV, the CXCR4-utilizing virus, wouldn't move in and take over. The Δ32 mutation couldn't protect against this. Hütter just had to wait and see what would happen inside Timothy's body. While infectious disease specialists warned him that the CXCR4 strain would take over, Hütter, an oncologist with almost no research experience, was convinced it wouldn't. It was an audacious act for a doctor who had never treated an HIV patient before. Unlike the infectious disease specialists, Hütter wasn't trying to enroll patients in a large-scale clinical trial. He, like Jessen, simply saw the opportunity to tailor a therapy to an individual.

But Hütter knew as well as anyone that Timothy's newfound resistance to HIV couldn't protect him from cancer. He might be able to fight off HIV in his bloodstream, but he couldn't keep the cancerous cells at bay. It was a cruel irony.

Two months after Timothy's cancer relapse at Christmas, he received a second stem cell transplant. The case was considered borderline; the oncologists were divided on whether or not he should have a second transplant. The chance of survival was about the same whether he received chemotherapy or a second stem cell transplant: 11 percent. Just like the first transplant, the second was a balancing act between getting the same HIV-resistant donor back in to donate and preparing Timothy for the transplant. Five days before the transplant, Timothy entered the hospital to begin the conditioning regimen, a course of many different drugs that suppress the immune system. The type of drugs given varies among doctors and hospitals, but the idea is to dampen down the immune system so that the patient and donor cells won't attack each other.

In addition to the drugs, Timothy received a single dose of whole-body irradiation. This is often given before a patient receives a stem cell transplant to clear out all the tumor cells as well as to help suppress the immune system.

All this left Timothy nauseated and tired. A few days before the stem cell transplant, the same anonymous donor started taking medication as well. These drugs, opposite to the ones Timothy was taking, spurred the donor's bone marrow into making more cells. He took the drugs by himself at home, getting ready for the big day. He didn't know Timothy, had no idea that he was part of a radical new idea to treat HIV, but he knew he was saving someone's life, for the second time.

After the second stem cell transplant, Timothy's recovery was very different. Unlike the rebounding health Timothy experienced after his first transplant, his health was now in steep decline. He had severe memory loss. He couldn't move his feet. Something had gone terribly wrong, and no one knew what happened. They performed a CT scan, to see if Timothy had an

internal injury following the surgery that could account for his
odd neurological symptoms, but the results were normal. What
they did find from their extensive tests were rapidly dividing
cells: a sure sign that the leukemia was back.

Hütter recalls that he then believed "that Timothy's chance
of survival was down to zero." He braced himself to explain the
situation to Timothy's family and friends. Hütter reckoned
Timothy had only weeks to live.

Meanwhile, Timothy had become delirious. Further analysis
of his tissue samples and those of his stem cell donor revealed a
shocking twist. The dividing cells that Hütter and the oncologists at Charité discovered, and subsequently used to diagnose
Timothy with a relapse of leukemia, turned out to be from the
anonymous donor. It was a shock to find that the donor specifically selected for his natural resistance to HIV was actually harboring cancerous cells not previously diagnosed. While the news
was disturbing for the donor, irony upon irony, it was good for
Timothy. Unlike HIV, leukemia isn't contagious; it can't be
passed from donor to recipient. Timothy's leukemia had not returned. But even though he didn't have a relapse, Timothy was
not getting better. If anything, he was getting worse.

The mystery deepened. Physicians could find no cause for
Timothy's bizarre set of neurological and physical symptoms,
which didn't match the side effects of a bone marrow transplant.
They tested Timothy for everything they could think of: HIV,
other viruses, bacteria, fungus. Nothing matched up. They began
to come up with elaborate theories: Perhaps decades of HIV infection had caused brain damage that was suddenly exacerbated?
Perhaps this was an unexpected side effect of total body

irradiation? While they debated, Timothy grew weaker in his hospital bed. To get to the bottom of the mystery, and hopefully save Timothy's life, the physicians at Charité hospital decided to perform biopsies of both his lymph nodes and his brain. Another operation. On his brain no less.

What they found was that during an earlier brain biopsy, performed shortly after his transplant, an accidental tear in the membrane that surrounds the brain had exposed the sensitive tissue to air and resulted in leakage of the cerebrospinal fluid into his body. This one tear accounted for 90 percent of Timothy's odd neurological symptoms. An operation was quickly performed to fix the tear in Timothy's brain.

By the end of 2008, Timothy was done for. He had been given a second stem cell transplant, had been told he had relapsing leukemia and would die, then was told it was all a mistake. He had a mysterious set of symptoms caused by a tear in his brain and had received three brain surgeries. After going through all this, Timothy was not himself; he was placed in the clinic for extreme brain injury, adjacent to Charité hospital. He lay in bed and watched endless hours of TV. He was incontinent. He couldn't tell the difference between his right and left legs. He was able to walk a little with the help of a walker but not far. His sight was dim. He was but a shadow of his former, dynamic self. For everyone around him, it seemed he wouldn't last long. Through it all, Lucas took care of his ex-boyfriend when many would have left. Timothy grew weaker. Timothy's mother had been visiting on and off in Berlin throughout the year but now, at the end, was staying for as long as necessary. Lucas, always considerate, let her stay in the apartment he had once shared with Timothy. One dark day, Lucas received the phone call he'd been dreading. "This is it," the doctor told him. "Timothy won't last much longer." Lucas began sobbing. Through the tears, he translated the news to Timothy's mother. Her reply was calm. She said, "I guess that's God's will." A very religious woman, she believed that Timothy's life was held in the hands of God alone. These words would wound Timothy and forever anger Lucas.

In any case, others would soon say God was smiling down on Timothy. His luck was about to change.

CHAPTER 19

"Perhaps We Have Eradicated HIV"

Christian had been off antiviral drugs for almost a year. It was October in Germany, a time of year when tourists come from all over the world to participate in the festivities of Oktoberfest. Berlin's streets were crowded with celebration. Christian, on the other hand, was nervous. For months now, he had been going in for blood tests, but today they were going to stick a needle in his lymph nodes to see if HIV was hiding in there.

For Jessen, the last few months had been surreal. Christian had kept all his appointments faithfully, coming in regularly to have his blood drawn. Christian said he felt healthy, he was sure the HIV was gone from his body. Remarkably, this seemed to be true. Despite using PCR tests sensitive enough to detect as little as 500 copies of the virus in his blood, they had been unable to detect any virus at all.

There were other signs that Christian had overcome HIV. In a healthy person, commander and storm trooper T cells are found in roughly equal numbers, in a 1:1 ratio. While a healthy person's

ratio ranges from 1 to 4, in AIDS patients, the number of commander T cells dips dangerously low to ratios below 0.5. This low
ratio means the immune system is in trouble. With dwindling
numbers of commander T cells, the immune system can't even
recognize which cells are infected with HIV, much less target
them for destruction. Physicians often use the ratio of commanders to storm troopers to assess the health of an HIV patient.

On the first day Christian took therapy, in June 1996, his
commander-to-storm-trooper ratio was 0.52. Having such a low
ratio early in the course of the infection shows that Christian's
immune system was already struggling. Surprisingly, this ratio
slowly climbed, even when he stopped his therapy early. Two
years after he first started therapy, and a year and a half after he
stopped, his ratio was 0.87, completely in the normal range for
a healthy person uninfected with HIV. As well as the improved
ratio, the sheer number of commander T cells had more than
doubled over the same period.

At the same time, the number of naive T cells doubled in
Christian's blood from a low 24 percent to a typical 49 percent.
Naive T cells are officers in training. They are the T cells that,
fresh from their time maturing in the thymus, are patrolling the
body, looking for invaders. Naive T cells stand in contrast to
memory T cells, which have encountered a foreign intruder and
remember it. These memory cells have been educated by war.
They subsequently become "activated," meaning they're ready to
plan an attack with the other cells of the immune system. The
fact that Christian's body had regained the naive T cell pool was
a welcome sign that the virus was no longer dominating his immune system.

When Jessen called Lisziewicz to tell her the virus was still undetectable in Christian months after he stopped therapy, at first she didn't believe him. She was sure there must be some mistake. Finally, as Jessen's insistent phone calls increased with intensity, Lisziewicz traveled to Germany. When she looked at Jessen's data, she remained incredulous. What he was showing her was impossible.

Then, timidly, she said aloud, "Perhaps we have eradicated HIV in this patient."

Both Jessen and Lisziewicz knew that if they wanted to prove that this one patient was truly cleared of the virus, they would need some heavy-hitting partners. They would need to bring in the big names in HIV that had both the methods and prestige to prove to the world that a patient had been cured. Lisziewicz's first call was to Bob Siliciano. He is an MD/PhD at Johns Hopkins University School of Medicine who, in 1997, published a highly influential paper in *Science*. Together with his colleagues, he had developed a new method to detect HIV in resting T cells. Resting T cells are just how they sound: They're inactive, unlike their dividing counterparts. Roughly 95 percent of the T cells in the blood are resting, awaiting the arrival of an outsider to propel them into action.

Because HIV likes to hide in resting T cells, measuring the amount of virus in these cells is a critical test in assessing the ability of any therapy to cure HIV. This is because clearing the virus from the blood isn't enough to cure someone of HIV. This became apparent in the mid-1990s when the new antiviral drugs proved themselves incredibly effective. Within a few months of using them, many patients went from high levels of HIV in the blood to none at all. Scientists hoped that the presence of these drugs was enough to wipe out the virus from the body so that patients wouldn't have to take the drugs forever. Siliciano's paper in 1997 dashed these hopes. His research showed that despite the fact that antiviral therapy drops HIV to undetectable levels in the blood, the virus is still hiding in resting T cells. These cells are a perfect hiding spot, for they can remain dormant for years, even decades, below the detection of the immune system and beyond the reach of antiviral therapy. The virus, stably inserted within our DNA, is able to bide its time, waiting until therapy is stopped and it can take over the immune system once again. Siliciano's research showed that the amount of virus hiding in these resting T cells does not decrease in proportion to the amount of time someone is on therapy. So it doesn't matter how long a person is on antiviral therapy—it can never wipe out the virus. At least not by itself.

This assay was obviously the one that Jessen and Lisziewicz needed to prove that their patient was different. That the unique therapy he had been given had cleared him of the virus. Siliciano hadn't had a patient with HIV yet that he couldn't detect virus in. This would be the ultimate challenge. Jessen sent half a liter of Christian's blood, about the same volume as a pint of milk, to Siliciano's lab in Baltimore and waited on tenterhooks.

Siliciano's group found something completely unprecedented. They couldn't detect any virus in Christian's blood. It looked just like the blood of a person who had never been infected with the virus. Of course everyone knew the experiment would have to be repeated and this was just a single result, but still . . . it was a wonder.

Jessen and Lisziewicz next needed to determine if there was any HIV in Christian's lymph nodes. Lymph nodes are tiny lima-bean-shaped organs distributed throughout our body. When we get a cold, we notice these tiny organs, often under the jaw, as they swell up uncomfortably, a sign that our immune system is up and running, fighting an infection. The lymph nodes act as a filtration system for the body, concentrating foreign particles. Millions of white blood cells are packed into each lymph node, the perfect staging ground for the start of an immune system attack. It's also the perfect environment for HIV to multiply and destroy. It's like a sneak attack on the enemy camp. When someone begins taking antiviral therapy, and the virus becomes undetectable in their blood, the virus can still be found lurking in the lymph nodes. Ultimately, because the virus multiplies to such high levels in the lymph nodes, it begins to destroy the organs, replacing the complex native architecture with a swath of scar tissue. This effectively cuts off the organ from the rest of the immune system. Jessen and Lisziewicz knew that if the lymph nodes were still intact, it would help explain the lack of virus in his body. They called Cecil Fox, a researcher, also in Maryland, who had just published an influential paper detecting HIV in lymph nodes. He was the leader in the field.

What Fox found when he examined Christian's lymph nodes was complicated. The lymph nodes were intact—a victory for the body against a virus skilled at destroying them. However, Fox could detect a "trace of HIV." While most research methods wouldn't have been able to detect any HIV in the lymph nodes, Fox, with his sophisticated equipment and extensive experience, was able to see something, although the amount was too low to quantify. Jessen and Lisziewicz, in no position to question an established expert in the field, decided that the best solution was to resample the lymph nodes. At the same time, they sent another half liter of Christian's blood to Siliciano's lab to repeat the search for HIV in resting T cells, the known reservoir of the virus.

Given the remarkable initial result, the Siliciano team redesigned their assay to make it five times more sensitive. They could now detect as little as a single infected memory T cell in a sea of 10 billion. The revolutionary technological feat was something of a compliment to Christian's immune system. This work paid off. Siliciano's group was able to find virus lurking in Christian's resting T cells, albeit at a very low frequency. Siliciano found that less than one of every billion cells in Christian harbored HIV. More important, they found that the virus hadn't changed. The lack of new mutations meant the virus had not been crippled by the immune system. Crippled virus is a phenomenon that would be documented later in HIV controllers, where the immune system put so much pressure on the virus that it mutated wildly to avoid the immune system attack. The heavily mutated viruses found in HIV controllers were no longer able to replicate and make more of themselves. They had effectively mutated themselves into a box, cornered by an immune system that was too smart for them. This wasn't so in Christian's case. The virus in him, when cultured in a dish in the incubator, was able to grow normally. So why wasn't it growing inside him?

The mystery deepened. A few months later, after the second set of Christian's lymph nodes was sampled, Fox found that only 3 of 4.4 billion cells harbored HIV. How was it that

Christian had these tiny, hardly quantifiable pockets of virus in hidden cells, but the virus hadn't surged back into his blood? Lisziewicz knew it must be due to some kind of immune response. She decided to make a call to the man best known for characterizing immune responses to HIV: Bruce Walker.

Walker was the researcher who, back in 1996, had published several papers looking at how storm trooper T cells target and destroy HIV-infected cells. He also had a small group of people who were infected with HIV but had no symptoms. Inside this small group of HIV controllers, he found something remarkable: Their storm trooper T cells were highly active against HIV. Walker developed an assay to measure the strength and specificity of these storm trooper cells against HIV. Lisziewicz knew this novel assay was the perfect tool for understanding how Christian was keeping the virus under control. If his body hadn't crippled the virus through a coordinated attack mediated by a genetic advantage, then perhaps the answer to this mystery lay in the unique therapy he had been given. Jessen and Lisziewicz hypothesized that giving Christian an intense, early therapy had beaten down the virus enough to let his immune system mount a proper attack.

When Walker got the call from Lisziewicz, he was shocked. This was just the case he had been waiting for. He believed that early, aggressive therapy was the answer, that it represented a path to eliminating the virus. He was simply waiting for the right clinical case to back up his theory and pave the way to new clinical trials. HIV researchers hardly ever use the word *cure*, a word that carries such import, it would be reckless to throw it around. Yet how else could one describe Christian's experience?

He had been infected with HIV, given an early, highly aggressive therapy, and no longer needed to take therapy. For all intents and purposes, the virus was cleared from his body. After speaking with Lisziewicz, Walker sent off a dozen e-mails to friends and collaborators. He was so excited about this fresh example of the power of HIV therapy, he couldn't wait to get ahold of some of Christian's storm trooper T cells, fresh from Berlin.

At the time, no company would risk shipping HIV-positive samples. Instead, Walker sent someone to fly to Berlin and pick up the precious cells. He sent Alicja Piechocka-Trocha, a technician who would work beside him his entire independent research career and who still manages the lab like clockwork. Returning the cells to Boston, Piechocka-Trocha prepared the assay Walker had developed when he was still in training, the enzyme-linked immunosorbent spot assay, nicknamed ELISPOT. ELISPOT, like the ELISA, is an HIV test that measures the ability of the immune system to recognize HIV and make antibodies against the virus. In this assay, instead of looking at antibodies, we measure the ability of the storm troopers to recognize and kill specific pieces of HIV. A clear plastic plate dimpled with ninety-six tiny wells was filled with tiny pieces of HIV, taken from each part of the virus. Christian's storm trooper T cells in varying concentrations were added to each well. Each condition was repeated several times. The storm trooper T cells kicked into action when coming into contact with a part of the gag gene, a key structural component of the virus that keeps the inner compartment of the virus intact. The cells released interferon-γ, or IFN-γ, a small protein called a cytokine. This humble protein is able to communicate with other cells, and it is a potent antiviral agent. IFN-γ is able to specifically recognize the double-stranded RNA of a virus and then draw in all the molecules and pathways needed to kill the infected cell. When Christian's cells released IFN-γ in response to a specific piece of the virus, it combined with a secondary antibody on the ELISPOT plate and turned the cells releasing the cytokine a bluish-purple. Each of these special wells became a polka-dot explosion, the number of purple dots indicating the intensity of the HIV-driven immune response. Piechocka-Trocha then put that plate in a reader that looks at each well individually and counts each tiny purple dot. For the part of HIV called gag, more than 2,000 cells released IFN-γ. It was a spectacularly vigorous response to the virus.

Finally, Jessen had an explanation for his seemingly miraculous patient. Christian's storm trooper T cells were able to mount an unusually powerful attack. It finally made sense how Christian could still harbor virus but the virus couldn't gain a foothold in his body. His immune system was able to keep the virus in check. Walker was elated at the news. This patient given early therapy now had an immune system that looked like one of his elite controllers. When talking to Lisziewicz about the data, he wasn't sure exactly what to call this Berlin patient. His name had never been given to Walker, to protect Christian's privacy. Instead, Walker decided to stay with "Berlin patient." It was a name that would stick, percolating down to scientists, to HIV support groups, and eventually to the press.

It's important to note that the ELISPOT assay is, like almost all assays, imperfect. It's impossible to replicate something as intricate as the human immune system in a well that holds less than a tenth of a teaspoon of liquid. You'll notice that nowhere in the description of the assay are mentioned a key player: the commanders. We also can't be sure how important IFN-γ is to the immune response to HIV. But despite these reservations, ELISPOT remains a powerful and frequently used tool to assess the strength of a patient's immune response to the virus. Clearly, this assay showed that, for Christian, his storm trooper T cells were able to recognize and respond to HIV infection in ways that most people can't.

Armed with a titillating case study, powerful data, and an all-star cast of HIV researchers, Jessen began to prepare the paper. He collected the data from the multiple collaborators as well as his own data from treating Christian. He wrote up a short paper and sent it off with the article to Lisziewicz, assuming he would be listed as first author. It was a reasonable assumption. After all, Christian was his patient, he was the one who decided to perform a small trial of the experimental medication inspired by Andrew, and he collated the data and wrote up the article.

In science, authorship is a precious and valued prize. The first author on a paper is usually the one who has done most of the work. The first author is typically the person who came up with the project idea, designed the experiments, and performed them. The first author nurtures the project like a baby, growing what was once only an idea into a real set of experiments, into a dataset that's analyzed, and into a paper that's published and finally read by scientists and journalists around the world. The last author, or senior author, is typically the person who funds the project. The senior author usually assists in interpreting the results and helps edit the paper. Between the first and senior authors is everyone else who worked on the project: technicians, graduate students, and collaborators. Even these names have a hierarchy, stretching down from those who put in the most effort to the least, with a special place reserved for "next to last" author, who often plays a significant senior-author-like role in the paper. These roles can shift; sometimes first and senior authors do more work and sometimes less. The hierarchy, however, is important. How many first- and senior-author papers a scientist has to his or her credit determines the ability to get faculty positions, to get tenure at a university, and to get grant money.

Everyone wants first author. For this paper, the situation was no different. Since it was being submitted to *The New England Journal of Medicine*, the competition was especially fierce. Being first author was already a big deal, but to be first author in such an important journal was a special opportunity. Immediately, the claws came out; everyone wanted first author: Lisziewicz, Walker, and, of course, Jessen. Remembering the fight that ensued over authorship, Lisziewicz calls it "sad." For her part, she felt she deserved first authorship at the time. She had coordinated the collaborators, bringing on the people needed to figure out what was going on in Christian's body. She was also the reason Jessen tested hydroxyurea in the first place. Walker feels bad about authorship on the paper as well, although his memory of the negotiations is limited.

Ultimately, Jessen was pushed out of the first-author position. It was given to Lisziewicz, and Jessen was shoved all the way down to fourth author. A surprising position for the scientist who decided to start a risky trial, recruited the patient, performed key experiments, and wrote up the paper. Some of those involved believe that putting Jessen as fourth author on the paper was blatantly unfair. One potential reason for his position on the paper is that Jessen was primarily a physician, not a scientist. Because of this, it was easier to claim that authorship was not as important to Jessen as it was to the other scientists involved. His salary comes from patients and insurance, not precious grant money.

Authorship settled, Christian's story was published in *The New England Journal of Medicine* in May 1999. The first line of the paper reads, "A patient, who has become known as 'the Berlin patient,' was treated soon after acute HIV infection." With those words, the tale of the Berlin patient would spread, seeping into research labs worldwide and fueling the imaginations of many living with the virus.

CHAPTER 20

An Unexciting Recovery

I n 2008, Timothy was coming back to life. When everyone was sure he would die—his doctors, friends, and family—he began to recover. He turned from a "vegetable" that lay in bed all day to a man who wanted to go for long walks. After the doctors repaired the torn membrane in his brain, he slowly returned to moderate activity. In a matter of weeks, he went from certain death to a rehab clinic. He kept to himself in the clinic, making no friends. Instead, he wandered around the hospital with his vision blurry and his legs weak. He found an Italian restaurant nearby that he liked, often eating there alone. Pieces of his normal life were coming back. Lucas, deeply relieved by his recovery, visited him, his new boyfriend in tow.

Life was moving on in a way. Timothy's memory was limited and his brain foggy. He had no job nor prospect of one. He felt lucky to be in Germany, where 26.7 percent of the gross domestic product is channeled back into the public welfare system; it's one of the most comprehensive welfare systems in the world. Compare this to the United States, which channels 15.9

percent of the GDP into welfare programs. Timothy relied on the small monthly stipend and free medical care he received from the German government to survive.

Hütter prepared to show other scientists Timothy's data. For the past year, since his first transplant, Hütter had organized a team of collaborators, all intent on analyzing Timothy's cells. In a show of solidarity, departments from all over the hospital donated their time and resources to this one-of-a-kind patient. They sequenced the virus hidden in his cells and measured the HIV-specific antibodies Timothy's body was making.

During this tense year, Hütter was inching closer to his hopes for Timothy. Each appointment brought the same news; the virus was undetectable and the CD4 T cells were steadily climbing. It was remarkable. However, the road still had some bumps. One of Hütter's biggest scares came five months after the first transplant, when a rectal biopsy was performed on Timothy. In the pinches of cells the biopsy had collected were CCR5-expressing macrophages. These findings were in direct contrast to what they found in the blood, where 100 percent of Timothy's cells were CCR5-negative and resistant to HIV. This was a bad sign. It meant that the strategy hadn't worked; they hadn't been able to replace all of Timothy's normal CCR5-expressing cells with the Δ32 mutant version the HIV-resistant donor had provided. Worst of all, these cells were located in the gut, HIV's breeding ground. Hütter was, again, unnerved. Oddly enough, the CD4 T cells in the same biopsy pinches were CCR5-negative.

He was about to get even worse news. Ultra-deep sequencing of Timothy's gut revealed the presence of viruses that use

CXCR4. Here was a disaster. The infectious disease specialists had told him this would happen. HIV would sneak around his lock on CCR5 and use a different receptor, CXCR4, instead. He waited patiently, fully expecting the CXCR4 virus to take over, even stronger than the original virus.

Though Hütter waited, the virus didn't come back. Neither did the new CXCR4-using virus take over. It didn't make sense. Was it possible Timothy was able to control the virus on his own? The cells in the rest of Timothy's body remained CCR5-negative, the virus was undetectable, and the CD4 T cells, which were close to zero in the days before the first stem cell transplant, were flourishing. They had slowly climbed to the normal, healthy levels of a person who has never had HIV. No doubt about it, Timothy was controlling what had once been decades of an insidious viral infection. Hütter's risky, unconventional experiment had worked.

He first presented the data at a small hematologist meeting. His data received no response. Hütter wasn't surprised. He had been expecting this. He knew hematologists like himself wouldn't be interested in Timothy's case. He had to bring the data to infectious disease doctors and HIV researchers. They, he knew, would not take his data so lightly. Excited, he applied for a talk at the Conference on Retroviruses and Opportunistic Infections, or CROI, one of the biggest meetings for HIV researchers. He knew his unique patient would capture the attention of those interested in new therapies.

At the same time, Hütter decided to write up his findings. He put together the data and wrote the manuscript. Just as with

Jessen's paper, a major battle over authorship ensued. Eckhard Thiel, the chief of transplantation medicine, although coming into the study late, took the senior-author position. He replaced Wolf Hofmann, originally considered the senior author. As compensation, the paper states "Drs. Hofmann and Thiel contributed equally to this article." The battle didn't end there. It was the first time the hospital would have a paper in *The New England Journal of Medicine*, and the power of being published in such a prestigious journal made everyone act a little nutty. All over the hospital, physicians were coming out of the woodwork, asking to be included on the paper. One physician who disliked Hütter challenged his first-authorship. This colleague even threatened to remove his name from the manuscript when he didn't get his way.

In the midst of this insanity, Hütter submitted the paper. It was an exciting period since it was the first time he had ever written a research paper. Not only was it his work, which he was immensely proud of, but he was submitting it to the top journal in his field. His excitement grew when he got the reviewer comments back. Scientific articles pass the first stage of acceptance when they are sent out for peer review. The article is sent to other experts in the field who anonymously cite the strengths and weaknesses of the paper and either recommend or reject it for publication in the journal. Though the reviewers exert significant interest, the decision ultimately rests with the journal's editor.

Although he didn't know it, Hütter was entering this process handicapped. One of the reasons Jessen's paper on the first Berlin patient was accepted so quickly was because of his collaborators.

Reviewers like to see names they recognize; it helps them trust the data they are evaluating. Hütter, with no prior record in HIV, and no collaborators in the field, was an anomaly. What's more, his data were provocative. There had never been a documented case like it. Speaking to the strength of the data, the reviewer comments were positive. The editor, however, was unimpressed. Hütter had no background in HIV research. He couldn't publish an article from an unknown entity; it was taking too much of a risk for the journal.

Hütter decided to resubmit his manuscript as a letter. This would mitigate the risk for the journal, for the responsibility for the contents of letters rests on that of the sender, not the publisher. In science journals, letters are not simply quickly jotted pieces of correspondence. They are highly crafted, peer-reviewed articles in their own right, and they carry significant prestige. Much to Hütter's dismay, the editor rejected his letter.

As this was happening, he learned from the organizers of CROI that he would not be awarded a talk. Instead, he would be allowed to present a poster of his data. Here was another blow. Poster sessions at CROI, while still valuable, are not nearly as prestigious as a talk. It showed how little the HIV community thought of Timothy's case. Hütter simply couldn't understand it. Here he had proof that a stem cell transplant from an HIV-resistant donor was able to turn the cells in Timothy's body into a lean, mean CCR5-negative machine able to keep HIV at bay. He had transformed Timothy from an HIV-positive person who took HIV medicine for a decade into a man who hadn't taken antiviral drugs in more than a year. How could the HIV community not be excited about this?

He traveled to Boston in 2008 for the CROI, carrying his poster entitled "Treatment of HIV-1 Infection by Allogeneic CCR5-Δ32/Δ32 Stem Cell Transplantation: A Promising Approach." His wording was careful; nowhere did he mention the word *cure*. Instead, he suggested that these results presented a therapeutic option for HIV-positive patients. It was a snowy afternoon in Boston as Hütter stood by his poster in the large conference hall. Right next to his poster was one from Bruce Walker and his collaborators in New York. Their poster, which was on reprogramming storm trooper T cells to make them HIV-specific, won a poster prize. Crowds gathered around the poster, full of eager, excited questions. Hütter's poster, on the other hand, remained quiet. It seemed no one was interested in his patient.

Just when it seemed his experience at the conference couldn't sink lower, Hütter attended a talk with some worrisome implications for his study. During the conference, trial results were released for a new drug: maraviroc. Maraviroc is designed to mimic the action of the Δ32 mutation. It sits protectively on top of the T cell, blocking HIV from using CCR5 to enter the cell. It is a similar philosophy, although a different approach, to Hütter's stem cell transplant. This particular study, because it looked at a skewed patient population, had some disappointing results. Hütter was stunned to learn that in 64 percent of HIV patients taking maraviroc, HIV switched from the common CCR5-using virus to a more aggressive virus that uses a different receptor: CXCR4. This was dangerous. Viruses that use CXCR4 cause a person to accelerate to AIDS faster. Hütter fretted over the implications for Timothy. Even if he succeeded in making Timothy resistant to the HIV he harbored, it seemed the virus would find another way.

Hütter's heart was heavy as he returned to Germany. Timothy might be dying in Berlin. The HIV community wasn't taking his research seriously. He couldn't publish his results. He was returning to a bitter hospital environment. In his hands, he held data proving that he had effectively cured a man of HIV. But he didn't know it yet. The future was bleak.

PART IV

The Cure

The more original a discovery, the more obvious it seems afterwards.
—Arthur Koestler, *The Act of Creation*

CHAPTER 21

Trials

H eiko Jessen was shocked at the headline. THE FIRST AIDS CURE? screamed across the page of *B.Z.*, a tabloid in Berlin. Inside, the pictures were ridiculous. They featured a man pretending to be the Berlin patient, a medical mask covering his face and a surgical cap pulled low. Pictures of Jessen treating a patient, as well as pictures from outside his clinic door, were featured prominently. The article went on to describe the case of Christian, describing him as the Berlin patient and pointing to his remarkable cure. Jessen was not amused. He had carefully avoided using the *cure* word and felt the tabloid had sensationalized the story. The past year had brought a blizzard of publicity for the young family doctor. He had spoken with countless news outlets, including *The New York Times* and *Newsweek*. It was this last interview that caused the most tension among his fellow authors of the *New England Journal of Medicine* paper.

Originally, the interview with *Newsweek* was supposed to be a full page, enough space to mention the many collaborators on the project and especially the new institute that Julianna

Lisziewicz and Franco Lori, senior author of the paper, had formed a few years earlier. New research institute funding was tight. Most institutes must rely heavily on donations from the private sector, beyond their portion of federal funding. With their Berlin patient, Lisziewicz and Lori had a nice opportunity to raise the profile and capital of their fledgling institute. They pressed Jessen to mention the institute during his interview with *Newsweek*, and Jessen was happy to oblige, mentioning all his collaborators, especially noting the role of Lori and Lisziewicz's institute.

Unfortunately, before the article went to press, there was a bigger news story afoot. The Kosovo War had gained momentum, and page space formerly dedicated to Jessen and the Berlin patient now had to be reallocated. What was a full-page article became a single paragraph, a paragraph that did not mention any collaborators. The effect was crushing on what had once been a close relationship. Conversations devolved into angry shouting matches. The friendship that Jessen and Lisziewicz had once shared, a friendship that had resulted in the Berlin patient, was permanently broken.

This fracture would have severe consequences for taking hydroxyurea therapy to the next level. The team was now on shaky ground going to clinical trial. In the wake of the *New England Journal of Medicine* paper, it seemed that each scientist had a different vision for how to translate the success of the Berlin patient into a viable therapy.

Bruce Walker believed the therapy itself was inconsequential; what mattered was its timing. If they could identify patients soon after becoming infected, before they displayed symptoms, and hit the virus with a heavy dose of antiviral drugs, they could potentially knock it down. Then when they stopped the drugs, the immune system would be at the ready to fight before the virus could get back on its feet. Walker's thinking on this came from a tiny group of acutely infected patients he and his colleague, Eric Rosenberg, were treating at Massachusetts General Hospital. They had been able to identify three people early in infection, before they displayed symptoms, and treated them aggressively with antiviral drugs. They drew blood from the patients, both before and after treatment, separated out the white blood cells, and stimulated them with purified pieces of HIV. They measured the proliferation of HIV-specific T cell responses, particularly the commanders. When they compared the data from these patients to elite controllers and patients infected with HIV for decades (chronic HIV), they found that the ability of the commander T cells to resist HIV in the acute group was as high as that of the elite controllers, both of which were far higher than chronically infected patients.

When they plotted this data and corresponded it to the amount of virus each person had in his blood, the data formed a perfect curve. The level of response from the commander T cells corresponded perfectly with the amount of virus. The higher the HIV-specific T cell response, the lower the virus. It made intuitive sense that the early therapy had somehow protected these critical cells necessary for the immune system to do its job. The data had some problems, though. Walker lacked a true control group; he hadn't found any newly diagnosed patient who didn't want to start therapy. Therefore, he couldn't directly compare acute patients who took therapy with those who didn't. Regardless, his data were convincing. He published his findings in *Science* in 1997. Even then, he didn't attribute much to the specific drugs given to each patient; they're not even mentioned in the paper. What was important was the timing.

The next step was obvious. They needed to stop therapy in patients who had been given this early, aggressive treatment. The problem was, there was no way to ethically do so. Walker knew that HIV patients taken off therapy could die. The researchers had no way to know if the strong T cell responses they were measuring were enough to protect the patients. Then the Berlin patient came along. He was the answer to their prayers. No one had documented an HIV patient in remission before. They now had direct proof that a patient identified early and given aggressive treatment, and who had stopped therapy, could control the virus. Even better, the Berlin patient's T cell responses were incredibly high. His T cells were clearly protecting him from the hidden virus lurking in his body. Walker could take his research to the next level; they would take patients treated in acute infection off therapy. They would, of course, watch them closely, making sure the virus didn't return. It could be done safely, he argued, as long as they were tested weekly for HIV. The minute the virus came back, they could be restarted on therapy. Walker was not alone. Other HIV researchers were pursuing similar lines of inquiry, all stemming from David Ho's original campaign for early therapy back in 1995. Where Walker was different was in his unparalleled ability to measure HIV-specific responses of the immune system and compare them to those special patients whose bodies controlled the virus without therapy. Now, with the Berlin patient, the approach made perfect sense; it was surefire.

Lisziewicz, Lori, and Jessen, on the other hand, believed that the hydroxyurea the Berlin patient had taken was the main reason why his body was now able to control the virus. Hydroxyurea was a unique drug. It targeted the ability of the cell to work instead of inhibiting HIV's enzymes. Hydroxyurea put a wrench in the machinery that created new building blocks of DNA, making a perfect space for fake building blocks, such as the drug didanosine, to infiltrate the viral genetic code. It also froze dividing cells, keeping the virus from establishing a foothold. The only drawback to hydroxyurea was its toxicity. Jessen felt strongly about the safety profile. For the trial in his clinic, he had used half the full dose recommended for cancer patients. Jessen had strong opinions on how a trial of hydroxyurea should be designed in order to minimize the toxicity of the drug.

Unfortunately, as the trial came to fruition, it was clear that Jessen's concerns would not be taken into account. Several small trials had mimicked Jessen's safety profile, each finding favorable results, although none included patients who had stopped therapy. It was impossible to know if the results would be similar to the Berlin patient's.

To know if the Berlin patient's experience could be replicated for other HIV patients, researchers needed to perform a large-scale trial of the drug, using a similar schedule of early treatment followed by what's called treatment interruption, a professional-sounding way of saying that patients stop taking therapy, "interrupting" their treatment. These interruptions would become known informally as drug holidays.

Performing such a large-scale trial is expensive. Lisziewicz and Lori partnered with the makers of hydroxyurea, Bristol-Myers Squibb, to get the necessary funding and infrastructure. As you can imagine, Bristol-Myers Squibb was very excited about the way the Berlin patient had catapulted hydroxyurea into the spotlight. They quickly began a trial through the Acute HIV Trials Group (ACTG), called ACTG 5025. The trial tested the same three drugs the Berlin patient received: hydroxyurea, didanosine, and indinavir. The doses, however, while the same, were given on a different schedule. Jessen had given Christian 400 milligrams of hydroxyurea three times daily, carefully balancing the doses with a strict eating schedule designed to promote absorption of the drug and limit toxicity, whereas the Bristol trial ignored these safety concerns and gave a single daily dose of 1200 milligrams. The reason for this was simple: It's difficult to adhere to a strict schedule as Christian did. Even more challenging is finding a large number of patients capable of following such a schedule. The antiviral drugs present enough of a challenge without adding another scheduled drug. If they enrolled patients incapable of taking the strict regimen, they risked getting no data from the trial. As it was, the trial ended up having worse consequences than a lack of data. It killed two people.

The trial enrolled 202 people, with a goal of 399. Volunteers were not newly infected with HIV. These were deemed too difficult to find and test, requiring a network of clinics like Jessen's. Instead, ACTG 5025 would recruit chronically infected patients already involved in clinical trials for antiviral drugs. Two patients in the hydroxyurea arm died of pancreatitis. Numerous toxicities were reported, including pancreas, liver, and nervous system damage. The trial was halted. Toxicity was now indelibly associated with hydroxyurea in HIV treatment.

Bristol, however, was not slowed down. At the 39th Inter-science Conference on Antimicrobial Agents and Chemotherapy (ICAAC) held in San Francisco in September 1999, the corporation used special sessions to promote the use of two of their branded hydroxyurea drugs for treatment of HIV. This flew in the face of FDA regulations that prohibit pharmaceutical companies from promoting off-label uses of their drugs. The labels on both of the company's hydroxyurea drugs specifically state that it's useful in treating several cancers, not HIV. At the meeting, they showed data from the Berlin patient, among others, stating that hydroxyurea had been proven as a first-line therapy for HIV. It was a bold, some might say outrageous, move considering that only four days earlier, they had been given notice that the ACTG 5025 trial was terminated. The corporation also listed suggested doses, including a 1200-milligram daily dose. They did not mention the two deaths that had occurred during their trial. In response, the FDA sent a warning letter, asking Bristol to stop using this language in their promotions, to send "Dear healthcare provider" letters to warn physicians of the potential danger of pancreatitis, and to bolster the warnings on the didanosine package insert.

Despite the mounting safety concerns over hydroxyurea, Lisziewicz, Lori, and Jessen were excited to take the lessons they had learned from the Berlin patient and translate them into a new clinical trial. To get enough funding , they turned to Bristol-Myers Squibb again. In particular, they would need the company to donate drugs for their study. As Lori, Lisziewicz, and Jessen were now pioneers in the use of hydroxyurea during HIV infection, receiving this support was not difficult. It did, however, come with a catch. The trial needed two Bristol drugs to replicate Christian's experience: hydroxyurea and didanosine. Bristol wanted to add a third drug they owned to the proposed clinical trial. This new drug was one that, unlike hydroxyurea, was new on the market. Jessen was convinced that this addition came because hydroxyurea had already been on the market for thirty years and so there were limited profits to be gained from it. A newly patented drug is able to make far more profit than one whose patent has already expired. Indeed, the new drug from Bristol, trademarked as Zerit, brought in sales of $605 million in 1999 alone.

But Zerit, despite its shiny new patent, was not really a new drug at all. In 1966, two years after Jerome Horwitz published data on his failed compound, AZT, he published the methods used to make a similar compound that he called d4T, designed to work the same way AZT worked. It imitates the DNA building block called thymidine. Like a ladder with a faulty rung, the drug sneakily incorporates itself into a growing chain of viral DNA but it's altered so that the next DNA base can't attach itself to it. It cuts off the virus from copying itself, thereby protecting more cells from becoming infected. Of course no one knew that d4T would make a potent antiviral drug until two pharmacology professors at Yale University, named William Prusoff and Tai-Shun Lin, dusted off the old compound and tested it against the virus in the early 1990s. The university patented use of d4T for treating HIV, then licensed the drug to Bristol. Bristol conducted several clinical trials of d4T, which was approved by the FDA as a new drug in 1994. The approval raised controversy since it was done by a special procedure for life-threatening diseases that allows a drug to be approved before it has been proven to work. At the time, Deborah Cotton, a Harvard professor who helped approve the drug, said, "I'm not sure how good our advice was today." She was referring not only to the effectiveness of the drug but also to its safety. Of the 10,000 people given d4T, 21 percent experienced neuropathy, a condition that causes pain and numbness, often in the hands and feet. Like AZT, d4T is toxic; the dose needed to be lowered to make it safer for people with HIV.

This is why, when Jessen learned that d4T, now trademarked as Zerit by Bristol, was being added to the clinical trial, he was not happy. "Here was a recipe for disaster," he recalls thinking. Hydroxyurea was already a highly toxic drug, but to add d4T was perilous. As Jessen looked around him, he realized there were few clinicians designing this trial. Who was looking after the patients' interests? He couldn't condone such an experiment. With a heavy heart, he left the group, disappointed at how things had fallen apart. It seemed to him that the collaborators he once stood in awe of had fallen prey to fiscal motivations.

It's surprising that following the Berlin patient, the first HIV patient to be in remission, no single investigator was trying to duplicate his unique therapy in a clinical trial. Instead, the treatment Christian had been given was broken up into two different clinical trials. Bruce Walker and his collaborators were testing one unique component of his treatment: giving antiviral drugs aggressively during acute HIV infection. Franco Lori and his collaborators were testing the other unique component: giving hydroxyurea, d4T, and didanosine to chronically infected HIV patients. It was timing versus aggression. In all the world of HIV research, no one was combining these unique components in the hope of replicating the Berlin patient's cure.

Sadly, neither of the early trials based on the first Berlin patient's therapy went well. Initially, Walker's data seemed remarkable. Published in *Nature* in 2000, Walker and his collaborators identified 16 HIV patients early in infection. All the patients were immediately started on antiviral drugs, most within 72 hours following diagnosis. The antiviral drugs did not include hydroxyurea. Twice because of hospitalizations, Christian had to go off therapy. Although they weren't planned, these two drug holidays, or treatment interruptions, were intriguing. Walker postulated that the interruptions were training the immune system to recognize the virus. That is, they gave the commander and storm trooper T cells an advance peek at the enemy. With this peek, they could tailor their attack for the virus. Then, once therapy was restarted, the cells were protected, ready for the next fight. If enough HIV-specific T cells were spared from enough rounds of treatment interruptions, they could provide a potent force against the virus. Those receiving treatment interruptions would then become like elite controllers—infected with HIV but able to control it. The strategy was aiming for a functional cure. Although the virus might not be eradicated in these patients, they would be able to live like Christian, free from having to take any antiviral drugs or worry about the virus.

Similar to how the Berlin patient underwent two treatment interruptions, 8 of Walker's subjects took one to two scheduled breaks. With the breaks scheduled, the trial was almost the opposite of David Ho's "hit hard, hit early" strategy, fashionable only a few years earlier. Of the 8 patients receiving treatment interruptions, 5 remained off therapy for an average of 2.7 years, with virus undetectable, at less than 500 copies per milliliter of blood. In addition, the HIV-specific T cell responses rose significantly. The results were impressive; the virus was not coming back. In his paper, Walker compared them to an acute HIV cohort not receiving antiviral therapy. In this group, similar to an experimental control, of 109 people, only 4 had less than 500 copies of virus per milliliter of blood at 2.5 years. The results fit perfectly with Walker's hypothesis—the HIV-specific T cell responses were high, and the virus was low.

The research immediately garnered media attention. Here, it seemed, was the answer to curing HIV. And the answer was so simple: Just stop therapy a few times. It was easy enough that anyone could do it. And people did. The popularity of the drug holiday rose—a well-deserved break from the strict schedules and handfuls of pills. Acute, chronic, young, and old all tried drug holidays, sometimes without telling their doctors. In 2001, Timothy Brown, then a translator in Berlin, took a drug holiday. He couldn't know that the holiday was inspired by the Berlin patient, a title he would eventually share.

The problem was that the drug holidays didn't work. In fact, they hurt. Following his drug holiday, the amount of commander T cells in Timothy's blood dropped to 250 cells per microliter, right on the edge of an AIDS diagnosis. The course of the virus varied from person to person, remaining hidden in some people for days, while in others, it lay quiet for weeks, months, even years. But the virus always returned. It turned out that even when the virus was hidden in the body, it was inflicting damage quietly. Those who took drug holidays experienced high levels of immune activation, a state in which the T cells, and often by-stander cells that just happened to be in the wrong place at the wrong time, are overly stimulated, causing them to die. Even worse, in some patients, the HIV strains they now harbored were resistant to antiviral drugs. Similar to how, if you have a bacterial infection and don't finish your antibiotics, you can develop antibiotic-resistant bacteria. In the HIV patients who took drug holidays, the virus had an advantage when reconfronted with an-tiviral drugs they had seen once before.

Treatment interruptions were a highly divisive issue in the HIV research community. In an interview in 1999, Anthony Fauci, head of the NIAID, was quick to question the safety of such interruptions, saying, "The strategy needs to be tested. The stop-and-go game can lead to drug resistance even if it looks so far like the wild type strain remains." Indeed, until the mid-2000s, investigators were still arguing bitterly over the benefits and risks of treatment interruption. Then, one study changed this: The SMART study began in 2002, enrolling patients in 33 countries. It was the largest study of its kind, enrolling 5,472 patients with a goal of 6,000 before it was abruptly stopped in 2006 due to safety concerns. The SMART (Strategic Manage-ment of Antiretroviral Therapy) study found that those HIV pa-tients receiving treatment interruptions were twice as likely to progress to AIDS. It was the final nail in the coffin of what had once been heralded as a new course for HIV therapy.

While Walker's trial of treatment interruptions in newly

infected patients soared and then plummeted, Lisziewicz and Lori were conducting a large-scale trial based on the other component of the Berlin patient's remission: hydroxyurea. Jessen had already left the trial in protest of the inclusion of d4T, a drug he believed was too toxic to be safely used in the study.

Unlike Walker, Lisziewicz and Lori believed that hydroxyurea played a key role in the Berlin patient's remission. Lisziewicz and Lori were invested in the drug's success, believing hydroxyurea was able to target the hidden reservoirs of HIV. Therefore, their clinical trial ignored other aspects of Christian's experience, such as treatment in acute infection and treatment interruptions, and focused on isolating the effect of hydroxyurea. The problem was that hydroxyurea had a tarnished reputation among researchers as well as patients.

The answer to this skepticism didn't lie in another Bristol-sponsored study, which, surprisingly, included the same drug dosage as the failed ACTG 5025. By the time Lori and Lisziewicz's study results were released, in 2005, hydroxyurea was a dirty word. This was especially unfortunate because Lori and Lisziewicz's study made important strides in determining a safe dose for hydroxyurea; they found that cutting the dose in half, from 1200 milligrams daily to 600 milligrams daily, reduced toxicity while maintaining similar levels of virus reduction and increased T cells. Sadly, their study was riddled with problems. Participants taking the higher dose of hydroxyurea, the exact same conditions as the failed ACTG 5025, experienced adverse events, including one death from pancreatitis, the same cause responsible for the two deaths in the earlier hydroxyurea study. For many reading their paper, only one line stood out: "The results of the RIGHT 702 study described here confirm that the use of high-dose hydroxyurea (1200 mg daily) can be associated with fatal pancreatitis." While other, smaller trials would pursue hydroxyurea as a therapy for HIV, none could surmount the reputation that hydroxyurea wasn't safe. In a field reluctant to publish negative results, we can never really get at the heart of hydroxyurea.

Lisziewicz believes the problem lies deeper than the safety issues embedded in those early clinical trials. She remains frustrated with the financial aspects of bringing a drug to the market. She believes it's too difficult to rouse interest in an older drug like hydroxyurea, which represents no major commercial interest and therefore no profit. Lisziewicz sums it up by saying, "If no one can make money, even the best drug in the world will fail." Even though they are no longer friends, Jessen agrees with her take on the drug that failed before its time, wishing that "it wasn't all about the money."

CHAPTER 22

Proof of Principle

The Conference on Retroviruses and Opportunistic Infections had been a difficult journey for Hütter as he struggled to establish himself as a researcher in the HIV community. Although his poster wasn't widely attended, he made some critical connections that would carry him on the path to publication. He met Steve Deeks, a physician and HIV researcher at the University of California, San Francisco, and Jeffrey Laurence, director of the Laboratory for AIDS Virus Research at Weill Cornell Medical College. Laurence, talking about Hütter's poster at the conference, said, "I thought it was the most exciting thing I'd heard about since the discovery of the virus. I couldn't believe people didn't take notice." Enticed by his work, Deeks and Laurence invited Hütter to a "think tank" sponsored by the Foundation for AIDS Research, or amFAR, later that year.

Returning to Boston in September, Hütter had a very different experience. He was no longer standing quietly next to an ignored poster in a crowded conference hall. He was now able to present his data to the HIV researchers in the community who could best appreciate it. The think tank discussed CCR5, reservoirs, eradication strategies, and viral latency. In 2008, it was a field that was blossoming, with new data pouring in from labs all over the world.

In the think tank, data were presented from John Zaia, a researcher at City of Hope hospital in Duarte, California. Zaia was pursuing a high-risk strategy in a small group of patients who had the same kind of cancer as Timothy: acute myeloid leukemia, or AML. Zaia was interested in a gene therapy approach to treating HIV. He was looking at no fewer than three different methods to knock out the CCR5 gene that HIV needs to enter human T cells. The first method built on work that Lisziewicz had done nearly two decades earlier during her time in the Gallo lab. Zaia used small pieces of HIV RNA, called short hairpin RNA, capable of binding to the virus as it tried to reproduce itself in a cell, so that the virus couldn't make more of itself. His gene therapy also included a decoy molecule, called a TAR decoy, to which HIV mistakenly binds when trying to insert itself into human DNA. Finally, Zaia's ambitious project also included a ribozyme, which is a uniquely structured RNA molecule that has an effect like an enzyme. The ribozyme he included was able to bind CCR5 on the body's cells, rearranging the atoms of the gene and blocking them from HIV's grasp.

He delivered these three very different gene therapy components using HIV itself, or at least a version of the virus crafted to be innocuous. The way most gene therapy works is that a virus takes genetic material into the body and circulates it. It might sound scary, but we can create viruses that are not harmful in themselves and, if they carry the right genes, can be extremely beneficial.

Zaia was testing this highly experimental gene therapy in patients who had both AML and HIV, just like Timothy. This patient population was ideal for his study, for they needed to
undergo the dangerous conditioning regimens necessary to receive hematopoietic stem cell transplants. Adding the gene therapy seemed like a small risk in comparison. There was another
reason this was an ideal population on which to test a gene therapy: It was a patient population with a high mortality rate, making it a group of people more likely to take big risks. Just as HIV
patients in the late 1980s were desperate for any clinical trial,
today AML patients with HIV have a high mortality rate and are
in desperate need of new interventions. Physicians and researchers are likewise torn about protecting the safety of their patients
versus giving them a chance at survival.

When John Zaia heard about Hütter's patient, he was stunned.
Here was the proof that his strategy could actually work. Even
though their methods were worlds apart, they were both targeting the same goal: Take down CCR5 to take down HIV. Zaia
knew that the existence of a patient such as Hütter described was
exactly the "proof of principle" that the HIV gene therapy field
needed to take his study seriously.

Just as Jessen couldn't have published the data from his Berlin patient without the backing of those hard-hitting HIV researchers who reviewed and bolstered his data, Hütter couldn't
get published without the help of the same major players in the
HIV world. At the top of the list was Bob Siliciano, the same
researcher who used his ultrasensitive HIV test to measure virus in resting T cells from the first Berlin patient; he now turned
his sophisticated techniques to Hütter's patient. Once again,

vials of cells and blood sera were being shipped from one patient in Berlin all over the world.

At the think tank, Hütter also met Mark Schoofs, a Pulitzer
Prize–winning journalist. Considering Schoofs was a journalist
and not a scientist, he would have a surprisingly important role
in Hütter's finally getting his paper published. In 1998, exactly
ten years earlier, Schoofs had interviewed Heiko Jessen, Bruce
Walker, and other key scientists on the first Berlin patient for *The
New York Times Magazine*. He had conducted the first of the only
two interviews with journalists that Christian would ever submit
to. Schoofs had played a key role in the subsequent media storm
over the first Berlin patient. Now, here he was talking to Hütter,
about to break the story of the second Berlin patient. Schoofs
wrote up the story in November 2008 for *The Wall Street Journal*
in an article entitled "A Doctor, a Mutation and a Potential Cure
for AIDS." When Hütter read the article, complete with his picture, he worried. Like Jessen before him, he hated seeing the
"*c*-word," *cure*, in the headline. He also worried that he had
stepped over a line. To go to the press before publishing in an
academic journal is a scientific sin. Those who fall prey to the
lure of media attention usually pay for it with rejection from
prominent journals. Hütter still hoped to be published in *The
New England Journal of Medicine*. Had he dashed his chances?

Data began to roll in from Hütter's new collaborators. Siliciano hadn't been able to find any traces of virus, and neither had
his other collaborators. The verdict was unanimous: Timothy had
a functional cure. When Hütter told Timothy, he was unimpressed. "But what about the cancer?" he asked. Being cured of
HIV was beside the point for Timothy.

Hütter revised his paper based on what he learned at the think tank and through the help of his new collaborators. However, the data in the paper didn't change. Hütter resubmitted his paper to *The New England Journal of Medicine*, this time, thankfully, getting a new editor and reviewers. He waded his way through thirty pages of reviewer comments, a process that was new and overwhelming to him. The reviewers nitpicked every piece of his data, sometimes, it seemed, willfully misunderstanding him. Despite the torture, Hütter answered all the reviewer comments himself.

The article Mark Schoofs wrote had the opposite effect than anticipated. Instead of making Hütter's actions seem self-aggrandizing, the article leant Hütter legitimacy in the eyes of the prestigious journal. His article was accepted and then published on February 12, 2009. The article itself, "Long-term Control of HIV by CCR5 Delta32/Delta32 Stem-Cell Transplantation," was a spectacular achievement. The Berlin patient who had been whispered about for almost a year in the HIV community was finally in print.

Hütter's paper begins: "A forty-year-old white man with newly diagnosed acute myeloid leukemia (FAB M4 sub-type, with normal cytogenetic features) presented to our hospital." Behind the clinical description stood a frightened man. Timothy, in the wake of the publication and subsequent publicity, didn't believe he was cured. He worried about his identity becoming known. He hated the idea of getting attention for his "cure" only to have the virus return. Timothy had been HIV-free for only two years, two very difficult years. A quiet, reserved person, Timothy couldn't imagine ever giving up his anonymity.

CHAPTER 23

The Good Doctor in Court

Heiko Jessen was experiencing a newfound fame in Germany. His patient load increased substantially. Everywhere he went, he was recognized. Simply walking down the street in his quiet neighborhood in Berlin, he was stopped by patients, friends, and admirers. His social life changed as well. Now when Jessen went out to bars and clubs at night, he found himself widely recognized. In some ways, it was great fun; the club bouncers shooed him in immediately, no cover charge, no lines. In other ways, it was uncomfortable, especially if he found himself surrounded by patients. This, of course, is the way of the family doctor. They are part of their communities and see their patients everywhere they go. What was different for Jessen was the new element of fame. He was no longer simply the sympathetic family doctor who treated gay men; he was now the famous researcher who could cure AIDS. Jessen had, in his words, "four excellent years." He had much success and for the first time, after going into considerable debt to start his practice, plenty of money. But Jessen's good fortune was about to change.

The tabloid *B.Z.* printed the Berlin patient's story under the headline AIDS DIE ERSTE HEILUNG? translated to English, "The First AIDS Cure?" Jessen had been careful to avoid the word *cure* in all media interviews. The effect was immediate. Hostility grew in his small medical community. In the wake of the sensationalized headline, numerous police reports were filed against Jessen. The claims were damning: They accused Jessen of cheating health insurance companies and income taxes, and receiving pharmaceutical and pharmacy money illegally. They claimed that Jessen was falsifying HIV diagnoses in an effort to increase his income, since he was paid more by insurance companies for HIV-positive patients.

In response, the police raided his clinic. They seized his medical records. They identified two hundred of his HIV-positive patients and invited them for HIV retesting. They went through his collaborators and medical colleagues to ensure that each record of additional testing, each extra exam, was real.

As the police sieved through his medical practice, they found a few errors that, in comparison to the larger charges, seem inconsequential. The largest of these mistakes was that Jessen was distributing methadone in his clinic to patients with drug addictions. Methadone is a dangerous drug given to help addicts wean themselves off heroin. The drug is as addictive as heroin itself, and regulation of it therefore must be tightly controlled. Physicians in Berlin can't simply prescribe methadone to patients who need it; they must have a special license for opioid addiction. The requirement for specialized training to prescribe methadone is not uniform: Some European countries allow generalist physicians to prescribe the drug and others require special licensing.

Jessen's case went to court. He was terrified; he could lose his license. Luckily, the sentence was minimal: a short probation and a fine for his oversight. The financial consequence on its own was small but was magnified by a poor investment Jessen had made in a housing development in the former East Berlin. The victim of an exploitive developer, Jessen found himself in a dire financial situation, and he went bankrupt. With the help of his brother, Arne, also a physician in Jessen's practice, he struggled through. He describes practicing medicine as a "sanctuary" during that time.

To make matters worse, Jessen's health was faltering. He noticed an odd pattern of red spots on his legs. He was having strange abdominal pain. After going to Charité hospital, he was diagnosed with an exceedingly rare disease: pneumatosis cystoides intestinalis. Gas had built up in his bowel wall. The disease was rare but could be fatal. It was 2002 and it seemed that everything in Jessen's life was working against him. As he received therapy at Charité, he decided he needed to take whatever small savings he had left and leave Berlin. He needed a long break from the city that seemed intent on punishing him.

Once Jessen learned he wouldn't die, he traveled. He spent two weeks in Sri Lanka, before leaving for a week in Singapore and then on to Dubai. It was just the escape he was looking for. After Andrew had left him, Jessen had been lonely. There was no one else out there quite like Andrew and he missed him. In Dubai he left behind old regrets and began dating someone new.

And his new boyfriend wasn't just anyone; he was a prince in Dubai. The romance was like a fairy tale to Jessen. His lover rode on a white horse, they spent their time together in his palace, and wherever he went, people were bowing to Jessen simply because they knew he was with the prince.

It was a life-affirming idyll. Jessen returned to Berlin rejuvenated, ready to dive into work and his research.

CHAPTER 24

Not Even Surprising

At an international workshop on HIV persistence and reservoirs, held in St. Martin in December 2009, Robert Gallo opened the conference with a challenge on the validity of Gero Hütter's report on his Berlin patient, Timothy Brown. It's not unusual for a newcomer in any scientific field to be met with skepticism. However, Hütter's shocking results and lack of background in HIV prompted an aggressive response. Gallo noted the absence of other, more established HIV researchers on Hütter's paper. He remarked that no one but a pathologist, examining samples from the patient lying dead on an exam table, could declare HIV was cured in such a patient.

Gallo's comments were on the minds of many in the audience. After all, this was not the first time they had been confronted with an extraordinary patient and the promise of a cure on the horizon. The original Berlin patient, Christian Hahn, had similarly garnered much excitement and been the subject of many promises. Early, aggressive therapy plus treatment interruptions had been touted as a cure in those few years before that therapy was debunked. The use of the word *cure* had been taboo ever since. No matter if Hütter wasn't using the *cure* word himself; he was implying it in his powerful data. In their defense, these researchers wanted to protect patients, to make sure that as scientists and physicians, they weren't raising false hope.

A dichotomy arises here. There is a protective nature in the bond that grows between patient and doctor. They know that the excitement that builds from the latest publication can evaporate when it reaches for the high bar of efficacy in real people. Researchers, on the other hand, are typically trained to minimize contact with the people behind their studies. Where physicians physically lay their hands on the skin of their patients, a researcher hides the patients behind a code of numbers and letters, obliterating all association to humanity. There is a good reason for this. Blind studies keep researchers from consciously or unconsciously influencing their observations. By keeping researchers and patients separate, the experiment has the best chance of escaping partiality and producing meaningful data. On the spectrum of physician and researcher, Hütter fit somewhere in the middle. He wasn't a family doctor like Jessen, but he also had little experience in research. He was simply a young physician/researcher trying to find the right balance between the two.

It was not fun for Hütter as he sat in the audience, listening to his research being censured by the preeminent HIV researcher. Sitting in his hotel room that night, he quickly amended his PowerPoint presentation, adding a slide titled "Do we have to cut this patient up in slices?" He delivered his talk the next afternoon to a much smaller audience than Gallo received for his opening remarks. However, that talk, "HIV Eradication by Stem Cell Transplantation: Is It Feasible?" was an eloquent and precise representation of his work. Hütter, intelligent and soft-spoken, was convincing. His credibility was raised after his talk, especially as other, senior HIV scientists lent their support. This small conference in the West Indies, so different from when Hütter presented his poster at the Conference on Retroviruses and Opportunistic Infections, served as an incubator of sorts, bringing his research to the HIV community in a way that made sense, even if no one knew how the Berlin patient's unique experience could ever be translated into a viable therapy for HIV patients.

When Anthony Fauci learned about Timothy Brown, just prior to Gallo's critical remarks, his reaction was mixed. As the director of NIAID, Fauci continues to have considerable sway in how the public and research community view the HIV field. He has to be skeptical when encountering new data since his opinion is so influential. Having lived through the dark days of HIV research, when newspapers and investigators alike were declaring the end of AIDS, Fauci knew firsthand the devastating effects of raising false hope. Therefore, when *The New York Times* asked him what he thought about the dramatic new case, his answer was in keeping with his skeptical personality: "It's very nice, and it's not even surprising, but it's just off the table of practicality."

It's a funny thing that sometimes when a result is spectacular, in retrospect the approach seems obvious. Now that the results of the Berlin patient's case were published and investigators everywhere were using the once forbidden *c*-word, the cure of HIV no longer seemed outlandish. Hütter's project had gone from one with data that no one believed to one that the head of NIAID called "not surprising." Hütter had to laugh at the absurdity of it.

But were they really "just off the table of practicality?" The aggressive nature of the therapy Timothy received, a bone marrow transplant, is not trivial. As we discussed earlier, it carries considerable risk. Bone marrow transplants have a significant mortality rate. Getting the transplant means going through much of what Timothy experienced: toxic therapy to make space in the bone marrow, the possibility of graft-versus-host disease, long hospital stays, and the potential for life-threatening complications. It's not a procedure anyone wants to sign up for. It's certainly not how we cure HIV for most people infected with the virus.

The other factor in considering any cure for HIV is the cost. A bone marrow transplant is one of the most expensive medical procedures, costing upwards of $300,000. A transplant such as the one Timothy received, using cells from a donor, costs approximately $805,400 in the United States. This number seems outrageously high until we compare it to the cost of a lifetime of antiviral therapy that runs $709,731 without discounts. This number, however, doesn't take into account other medical expenses that HIV patients have. As people are living longer with HIV, their medical complications and therefore their expenses are also on the rise. In the United States, survival has increased from 10 years in 1996 to 22 years in 2005. With that increased survival rate, other problems have arisen. Those with HIV are more likely to suffer from premature aging, incurring significant medical bills. Statisticians take all these factors into account when they compare therapies by QALYs, or quality-adjusted life years. QALYs measure both the quantity and quality of life gained by a medical intervention. QALYs are then used by health insurance companies, hospitals, and even nonprofits to determine cost analysis, essentially whether a medical intervention is "worth it." Any new HIV therapy is measured using this system to determine whether it makes financial sense. By any measure, the therapy Timothy received does not make medical or financial sense for an HIV-positive person not already in need of a bone marrow transplant.

But what does make practical sense is to take the cure we've gleaned from Timothy and turn it into a therapy that is both medically and fiscally responsible.

Both medical and fiscal concerns are why we need a practical cure. Atripla, a popular antiretroviral drug that combines three separate medications in one pill, costs about $20,000 a year. This causes problems in the United States, where cash-strapped states have trouble covering the cost through Medicare programs, and private insurance often caps yearly expenses. In an era when HIV-positive individuals have more choices than ever before, cost can still be a limiting factor.

In addition to cost, not all patients are able to find a therapy that works for them. Jason, who has been HIV-positive since 1988, gets angry when people he knows tell him that the fight for a cure to HIV is irrelevant. "They simply don't know what they're talking about," he grumbles as he sits in his family's house in Petaluma, California. "I've spent decades trying to take medications I can't tolerate, only to see my immune system crumble." His longtime partner, Richard, agrees, saying, "The gay community doesn't care about HIV anymore." Even with the best medical help, Jason can't find a therapy that restores his T cells. He is constantly battling illness as a result. As we stand in Mission Dolores Park in the Mission District of San Francisco, you can almost see the anger, mixed with grief, rising off him. In the wide-ranging world of antiviral drugs, we sometimes forget that there are people like Jason and Richard out there, still struggling.

HIV is more manageable today, but that doesn't mean it's easy to live with. Not everyone can find a therapy that they can tolerate, that is effective, and that they can afford. Even if they do find a therapy, those with HIV face shortened life spans. The average life expectancy is just fifty-eight years for men, compared to seventy-three for men who aren't infected with HIV. In other countries, life expectancy is even lower. It isn't just a matter of age. Dementia, arthritis, and other diseases of our nervous system are all found at higher rates in HIV-positive individuals. This is because HIV is able to creep into our brains. In fact, more than 40 percent of HIV-infected patients have some kind of neurological condition.

While the virus can cross the blood–brain barrier, the seal that separates our central nervous system from the blood circulating around our body, the drugs we use to target the virus cannot. Because of this, HIV is able to outrun the therapy gunning for it, and infect and replicate in the brain. A whole new virus population forms in the brain, genetically distinct from that in the blood. This new world of viruses can cause inflammation and cell death and is linked to HIV-associated dementia.

Aging and HIV is a relatively new field in HIV research. Many researchers believe that we're just at the tip of the iceberg in learning how the virus accelerates aging. With so much research focused on prevention, few resources have looked at how to manage the effects of HIV in people over fifty. Yet as patients survive longer on new classes of antiviral drugs, this is a growing population.

HIV exhausts the immune system. Each cell in our bodies has a built-in death clock, a limited number of divisions it can make before it peters out and dies. HIV stimulates the immune system to divide at a frantic pace as it rushes to provide enough cells to target the virus. This is why some scientists think, if therapy is started earlier, before the virus decimates the T cell population, life expectancy will dramatically rise. While still controversial, some reports have shown, on average, over a decade is gained from early therapy.

If we have enough money, if we can find a combination of antiviral drugs we can live with, and if we start early, we can live a long time with HIV. The problem is that these three criteria are not always easy to come by. Our background and our geography all present hurdles, hurdles that many of us cannot jump. This is why it's not enough simply to keep producing new HIV drugs, to stay ahead of the ever-mutating virus. We need a functional cure.

Large-scale clinical trials building on Christian's success were a failure. These trials all captured a piece of Christian's therapy, but none of them replicated his experience. The reasons for this are manifold. On one hand, it's difficult to recruit people recently infected with HIV. Many people aren't like Christian and don't know the date of their exposure so precisely. Even if they do, they may be loath to go to the doctor so soon. The reason Christian was able to get such early therapy was his relationship with Jessen. He had a family doctor he trusted and was close to. Not many young people can say the same. Particularly not in the United States, where the large-scale clinical trials were held. Young men aged 15–24 are the least likely in the United States to regularly see a primary care physician. Even if they suspect they're infected, they usually have no relationship with a medical health professional to help them navigate the waters. This is a major stumbling block for clinical trials looking at early therapy. It's difficult to recruit patients diagnosed quickly after infection and who are at similar stages of the disease. Because there was so much variation among patients, the trials could conclude nothing; they couldn't even recommend when to start therapy. It was tremendously frustrating.

A catch-22 had been established. The only way to get people recently infected with HIV to seek therapy early was to have strong scientific evidence that early therapy could help. But the only way to get that evidence was to persuade more people to come in early after becoming infected with HIV.

Christian hadn't just received early therapy; he had also already received a potent cancer drug. Unfortunately, because none of the large-scale clinical trials replicated his experience, it's impossible to know the influence of each individual component in contributing to Christian's cure. Unpredictably, a similar problem exists with Timothy's cure. While it's likely that Timothy's cure was mediated by the genetic advantage given to him by the HIV-resistant mutation Δ32, Timothy also received a mélange of other treatments that might have influenced his cure. These included conditioning therapy, graft-versus-host disease, and the stem cell transplant itself. It was impossible to weight the influences of these factors in his cure. In both Berlin patient cases, you had an unusual finding: Doctors had cured a patient of HIV, but scientists were still debating how that cure happened.

Both Christian's and Timothy's cases were made more complicated by their individual genetics. Timothy was heterozygous for the Δ32 mutation before he had the transplant. He had only one functional copy of the CCR5 gene rather than the two functional copies most people have. This meant he had a genetic head start. How had this genetic advantage influenced his cure? It was impossible to measure. Christian also had an interesting makeup in his HLA, the genes that govern the immune system. It turned out that Christian's HLA type was B*57.

People with certain subtypes of this HLA are more likely to be the special elite controllers of HIV. However, while this HLA type is seen more commonly in people who genetically control the virus, the vast majority of people with this HLA cannot control HIV. Moreover, Christian was negative for a subtype of B*57, the B*5701 gene, most commonly associated with elite controllers. Christian's symptoms did not match those of an elite controller or even of someone slow to progress to AIDS, called a slow progressor. That's because Christian, unlike the genetically gifted elite controllers, had high levels of virus after becoming infected with HIV, suffered from infections due to his weakened immune system, went on antiviral therapy, and experienced an increase in virus when he went off the therapy. He also didn't harbor the crippled virus that HIV controllers have, which indicate the ability of their immune systems to mangle the virus without the help of drugs. Nevertheless, the fact that Christian has the HLA-B*57 gene created confusion.

For the physicians intimately involved with the Berlin patients' treatment, the controversy stirred by the individual genetics of the patients seemed silly. Both Jessen and Lisziewicz believe that Christian's cure came from the therapy he was given, combined with the early timing of the therapy. However, other researchers have postulated his control came from his individual genetics. Without a follow-up study, we may never be able to completely rule out his HLA-B*57 gene as a factor in his cure. Similarly, Hütter believes that Timothy's cure came from the HIV-resistant cells he was given. Just as in Christian's cure, other researchers question this. They believe Timothy's cure may come from the myriad other medical aspects involved in his therapy. One thing we can count on: A spectacular medical finding is scrutinized to the very last detail of doubt.

For both Berlin patients, their individual genetics seemed to be getting in the way of what should be obvious: Both men had HIV and both men were cured. Both cases were mired in controversy and scientific debate. But at its heart, both therapies represented a new way of thinking about curing HIV and embodied a new strategy to get to that cure. The question was how to use what scientists had learned from these individual cases to inform new strategies. Ultimately, while both Berlin patients were cured of HIV, it wasn't a cure that anyone would want.

Scientists would have to take their inspiration from these cases and find a way to translate the ideas into therapies for everyone. By themselves, the Berlin patient cases were anomalies. The Berlin patients were not the answer. They were invitations to fulfill a promise, the promise that HIV can be cured.

CHAPTER 25

The Promise Kept

As the video opens, we see Emily Whitehead, only six years old, lying in a hospital bed. Emma, as she likes to be called, is wearing purple, her favorite color. Her head is bare, the multiple rounds of chemotherapy having stripped her of her once thick, brown hair. She sits emotionless, watching calmly as doctors quietly explain what each tube being placed in her body is for. In 2010, when Emma was five years old, she was diagnosed with acute lymphoblastic leukemia, or ALL. This kind of leukemia is a close cousin to the one Timothy had. Just as the storm trooper T cells in a person with HIV have trouble hunting all the cells infected with the virus, in leukemia, the storm troopers have trouble hunting all the cancerous cells.

Emma went through a year of chemotherapy before learning that her leukemia had reappeared. The relapse was a terrible sign, reducing her chances of defeating the cancer from 80 to 90 percent to less than 30 percent. She started aggressive chemotherapy treatments and was scheduled for a bone marrow transplant in February 2012. Exactly like Timothy's procedure, the bone marrow transplant would boost her immune system, giving her all the precious immune cells the cancer was killing off. But just two weeks before the transplant, Emma learned she couldn't receive it. She had again relapsed. Emma went back on chemotherapy as her parents hoped for a miracle. The options for her were shrinking. Cells taken from her bone marrow with a big needle placed in Emma's spine, showed that 7 percent of the cells were cancerous. The chemo wasn't working. Emma and her parents were left with one option: a highly experimental clinical trial at the Children's Hospital of Philadelphia. They had rejected the idea of this clinical trial once already. It was a frightening procedure, and Emma would be the very first child to participate in it.

The clinical trial, called CART-19, would isolate the T cells from Emma's blood and then genetically engineer them to specifically recognize and kill the cancer cells lurking in her body. How can we transform T cells into cancer-killing machines? The trick, as mentioned above, involves HIV. Carl June, the lead investigator of the trial, knew that HIV is a master at getting into cells. In order to safely exploit this feature of the virus, June used a version of HIV that had been spliced apart, removing all the parts of the virus that make it dangerous. It's the same vehicle, or vector, that John Zaia, working three thousand miles away in California, was using for his gene therapy. June then stuck the cancer-targeting information the cell needed into the empty shell of the virus. HIV worked as the packaging for the information the cell needed to target cancer. Like a sheep in wolf's clothing, the virus was able to invade the T cells. Once inside, instead of taking over the cell's machinery to replicate itself, the virus delivered the blueprint for destroying cancer cells. That blueprint was a chimeric antigen receptor, or CAR. The CAR is a modified version of the T cell receptor, or TCR, the molecule that lies on the surface of T cells and is a key part of how we control our immune system. By modifying the TCR, researchers were changing how the immune system responded to invaders. The CAR that Carl June is using is composed of signaling molecules that lie on the surface of B cells. The gene therapy replaces how T cells decide which target to kill, so that they now put all their energy toward attacking the B cell and B-cell progenitors in the bone marrow (and the rest of the body) that are harboring the cancer.

On April 17, 2012, Emily became the first pediatric patient to receive CART-19. Her T cells were isolated from her blood and then treated with the tempered HIV containing the cancer-killing blueprint. Over the course of three days, those T cells were reinfused into her body. It would take ten long days to know if the T cells were doing their job and killing off the cancer. However, after only three days, Emma became very sick. She was running a fever of 104.9°F, and became delirious. She was rushed to the pediatric ICU, where her breathing tanked and her blood pressure became dangerously low. In a last-ditch effort, she was given steroids, a move that physicians knew could kill off the genetically engineered T cells. The gene therapy was now unimportant as Emma's life hung in the balance.

In Ross Kauffman's short video "Fire with Fire," Carl June describes what happened next: "It was like the calm after the storm. The clouds went away. And she woke up and there was no leukemia." With tears flooding his eyes and a shaky voice, June says, "When that child survived it was a course. An amazing event." Emma survived and her cancer is in remission. Today, she is a beautiful little girl with a head of brown, wavy hair, and still loves purple.

The gene therapy Emma received was informed by research trying to replicate Timothy's cure. Timothy's experience has influenced not just the path of HIV cure research but also that of cancer gene therapies. This is because Carl June was able to refine Emma's therapy based on lessons learned from modifying T cells in HIV gene therapy trials targeting CCR5.

Carl June's draft number was 50. It wasn't a good number. Any American man born between 1944 and 1950 with a draft number less than 195 was classified as 1-A and required to report for service. June knew what that number meant, that he would be fighting in the frontlines of the Vietnam War. It was 1971 and he had just graduated from high school. Two years later, the war was over, but June's career in the military was just beginning. He went into the United States Naval Academy in Annapolis, and then, in an effort to "control his destiny," June decided to go to medical school. It was a way to exert control over what the long years of his military obligation would be like. By the early 1980s, he was among a small group of physicians chosen to learn a complicated procedure in Japan: bone marrow transplantation.

For the military, this move was not about treating cancer. In the 1950s, as fears of nuclear warfare grew, the military considered radiation poisoning to be a considerable threat to the public. During the Manhattan Project, a group of scientists observed that the spleen seemed to offer protection against radiation poisoning. Building on these observations, in 1951 they tested the first bone marrow transplant in mice. Their results were remarkable: The procedure, infusing the stem cells taken from the bone marrow back into the mouse, was able to rescue the animal from a lethal dose of radiation.

June would learn bone marrow transplantation from the master: E. Donnall Thomas, the man who performed the first bone marrow transplant in a human in 1956. His work, which drastically improved the survival of those with blood cancers like Timothy's, won him the Nobel Prize in Medicine in 1990. With June's interest in cancer research now piqued, he was disappointed to learn that the program in Japan was being discontinued. There was simply no cancer research being done in the military. And since he couldn't work on cancer, he decided to work on an infectious disease program in which the Navy was beginning to invest heavily: HIV. Although he would never have chosen such diverse training in cancer and HIV if left to his own devices, thanks to the military, June had a perfect background for developing a completely novel approach to both diseases.

In the mid-1990s, June was working at what is now the Walter Reed National Military Medical Center in Bethesda, Maryland, across the street from the NIH, when a new postdoctoral fellow came to interview for a position in his lab. Bruce Levine, from a family of scientists, had been working in labs since high school. As he interviewed for June's lab, he sat two floors above the maternity ward where he first entered the world. It felt like a sign. Levine joined June's lab, excited as a PhD to be part of the hospital environment.

Levine and June began looking at how to make T cells grow outside the body. At the time, growing T cells in the lab was a major challenge, involving either a complicated mix of cell-signaling molecules or the use of dendritic cells. As you can imagine, this was not good for those researching HIV. Researchers needed a simple way to model the virus in human cells. June and Levine took on this problem by creating an artificial dendritic cell. Dendritic cells are immune cells formed all over the body. They're called dendritic because of their funny, treelike appearance, where the edges of the cell branch out like roots of a tree. In addition to other functions, they provide signals to the T cells that tell them to mature. June and Levine developed an artificial dendritic cell, a tiny magnetic bead that could induce the same effect in T cells. The artificial cells were a success; by adding them to the cultured T cells biweekly, the T cells could be grown in the incubator easily. But something odd happened when they tested their technique in T cells taken from HIV-positive individuals. The T cells that once harbored virus were suddenly resistant to HIV infection. Here was a mystery: How were the artificial dendritic cells able to confer resistance to HIV infection?

It wasn't until after they published their paper in 1996 that the answer became clear: The T cells growing in the incubator didn't express CCR5. The artificial dendritic cells, in addition to telling the T cells to mature, were also clearing CCR5 off the cell surface. Without CCR5, the cells couldn't be infected with HIV.

These observations stuck with June and Levine as they continued their research. Moving their lab to the University of Pennsylvania, in 2004, June received a visit from an old friend: Dale Ando. Ando had bounced around several different biotech companies before landing at Sangamo BioSciences that year. He came to June with a crazy idea: a "Star Wars approach" to treating HIV. "What if," Ando postulated, they "could knock out the co-receptor HIV needs to enter T cells?" The idea sounded crazy. Even more crazy was the data Ando had. Their efficacy at knocking out co-receptors was less than 1 percent. It was madness to spend time and money on such an inefficient technology. If it had been anyone else but his friend Ando, June would have likely dismissed the whole concept. As it was, he told Levine about the idea and then added dismissively, "Yeah, right, like that's going to work."

Edward Lanphier founded Sangamo BioSciences in 1995. Intensely interested in gene therapy, Lanphier had been working for Somatix, one of a host of start-ups that had jumped on the gene therapy bandwagon. It was a challenging time for gene therapy. Almost every gene Lanphier wanted to acquire was already owned. Restrictions based on intellectual property were weakening the new industry. Lanphier says that "it was not optimal; you would get what you could get."

In the midst of negotiating these complicated deals, Lanphier became interested in the work of Srinivasan Chandrasegaran. Chandra, as his friends call him, was a postdoc in Jeremy Berg's lab at Johns Hopkins University when he created and patented zinc finger nucleases (ZFNs), which are small gene-editing machines. To make them, Chandra put together two mechanisms, both borrowed from the natural world. The first are zinc fingers. Zinc fingers were first discovered when looking at the RNA of the African clawed frog. Scientists wondered how the RNA of this creature was able to bind both strongly and specifically to a protein. They discovered that the secret was due to the unique structure of a protein, which had elongated fingerlike structures held together in the center by a zinc ion. Here, in nature, was a perfect example of how to specifically target and grab a piece of DNA: a zinc finger. Chandra stitched together several zinc finger proteins and attached them to an enzyme that cuts DNA. These enzymes, called restriction enzymes, were first discovered in bacteria. Amusingly, the bacteria use the enzymes to fight off viruses. The enzymes cut out the intruder DNA from the native DNA. The enzymes are a powerful tool in molecular biology and cloning, allowing scientists to cut up the DNA they're working with and rearrange it. By combining the DNA-gripping nature of a zinc finger with the DNA-cutting ability of a restriction enzyme, Chandra had created a novel tool.

Somatix, however, wasn't interested in zinc finger nucleases. That didn't deter Lanphier, who knew he had found a powerful new technology. After being tangled in the intellectual property laws that were crippling gene therapy, Lanphier appreciated this fresh approach to modifying genes. He decided to take a chance and form his own company. Going to family and friends, he raised $750,000 to start the business. When remembering those years, Lanphier says he "should have been scared." But instead those early years were exhilarating.

The idea to tackle HIV would come years later when Dale Ando joined the company. Between him and Philip Gregory, the chief scientific officer at Sangamo, they devised a plot to use the specificity of the zinc finger nucleases to attack the co-receptor HIV needs to enter T cells.

Each zinc finger is designed to bind only one specific part of a gene. The zinc finger nucleases Sangamo designed for HIV are specifically created to target CCR5 and only CCR5. The zinc finger matches up to twelve building blocks of DNA, the A, T, G, and C that make up the DNA alphabet. To knock out the CCR5 gene, it's not enough to take out one strand of DNA, which the cell can repair. Instead, both strands of DNA that encode CCR5 have to be cut because a double-stranded break is not repaired well by the cell. Our repair enzymes need the information contained in the complementary strand in order to reconstruct the gene. It's similar to when a building is demolished, if only one wall is taken down, matching up the structural work to the remaining walls can likely repair it. But if we take down all the walls, the building is done for.

For this reason, two zinc finger nucleases are delivered to the nucleus of the cell. Each zinc finger makes its way to its specific target: the single strands of DNA that encode the CCR5 gene (see the illustration on page 230). The zinc fingers bind to the DNA, gripping the molecule strongly. Only after the DNA is in its grasp does the nuclease portion of the ZFN engage. Alone, each nuclease doesn't have the ability to cut DNA. But, when two ZFNs line up perfectly, they form a dimer, two halves making a whole, creating a precise break in both strands of DNA. Here was the way to tackle HIV. Using the CCR5-specific ZFNs, they

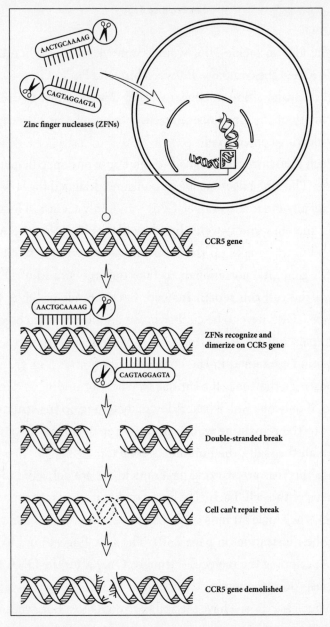

How zinc finger nucleases (ZFNs) take out CCR5. Two ZFNs are delivered to a cell. Each contains a region that binds to a part of the CCR5 gene. When they are both attached to the gene, their accompanying restriction enzymes combine and cause both DNA strands to break. Unable to make the CCR5 protein, the ZFN-treated cell locks out HIV.

could wipe out expression of CCR5 on the surface of T cells, keeping HIV from entering the cells.

The gene therapy could potentially work as both a vaccine, keeping healthy people exposed to the virus from ever becoming infected, or as a cure, eradicating the virus from the body of those harboring the virus. It was a highly original approach, perhaps too original. For most scientists hearing it for the first time, it seemed crazy.

Carl June was no exception. Nevertheless, by the twists of fate, June had a background tailored to pursue the unusual project. He had experience working with CCR5 and HIV from his days in the 1990s culturing T cells. He was actively pursuing gene therapy for cancer patients and was interested in how it could be applied to HIV. He had experience in engineering and transplanting blood cells. All these components combined to give him the exact experience needed to tackle such a seemingly crazy and complex project. In fact, it would be hard to imagine any other researcher able to grapple with the amalgam of research experience needed.

June and his lab grudgingly took the CCR5 ZFNs from Sangamo and began testing them in T cells taken directly from patients. Behind closed doors, they laughed at the method, calling it the Star Wars approach for HIV. Working with the finicky human cells, they were able to optimize the disruption of CCR5. They then took the cells and injected them in a mouse model. But it wasn't just any mouse model. June and his lab chose a humanized mouse model.

Humanized mice are the latest trend to hit the animal model world. The problem with modeling disease in animals is that it can never completely mimic pathogens in humans. Human diseases simply behave differently in a mouse than they do in a human. For HIV, the problem is severe. Mice can't be infected with HIV. So instead, we turn to monkey models of infection. But even here, in our primate cousins, we run into problems. Monkeys can't be infected with HIV, with the exception of chimpanzees, an animal model no longer used because of their endangered status in the wild. Instead, the monkey models we use are infected with SIV, or simian immunodeficiency virus. SIV, a close relative of HIV, behaves a lot like its human counterpart, but is certainly not identical. There are over 40 strains of SIV, native to different primate species. Despite the diversity of strains to choose from, genetically, only 50 percent of the SIV strain most commonly used in research studies matches that of HIV.

Perhaps the biggest difference between HIV and SIV is that most SIVs don't cause disease. Monkeys live out their lives with the virus multiplying inside them with little effect. The virus and monkey coexistence has been tweaked over time, evolution smoothing the path to a stable truce between virus and host. The reason monkeys have this and we don't is because they've lived with SIV longer than we have, likely over 32,000 years. Compare this to HIV, which has had only one hundred years to adapt to us. SIV has had the time to work out the kinks.

For this reason, most SIV models aren't very good for studying HIV. Thus, researchers developed SIVmac, an SIV taken from a sooty mangabey and put into a rhesus macaque monkey. Rhesus macaques are a species of monkey that aren't naturally infected with SIV in the wild, so they haven't had time to adapt to the virus, especially a virus from a different monkey species. This is why SIVmac behaves more like HIV: It causes disease, even death.

All the HIV vaccines that have made their way to clinical trials have done so through the SIVmac model. The path to these unsuccessful vaccines, although numerous factors resulted in their failure, was paved in macaque monkeys. Although the vaccines were able to protect the monkeys from getting SIV, researchers were unable to translate this success to humans. There are simply too many differences between the two viruses and between monkeys and us.

In humans, it typically takes between seven to ten years to develop AIDS. In the SIVmac model, AIDS is usually reached within six months. Extreme levels of virus and deteriorating CD4+ T cells characterize these turbulent months. In comparing the disease courses of the two, SIVmac bears only a resemblance to HIV.

Other problems that arise from using the monkey model are its expense—housing and experimentation on primates requires large primate centers with ample funding—and the sheer impracticality of obtaining enough monkeys. This is a sticking point in science, where statistics hinge on the size of a study. That said, SIVmac was the only game in town. It might not be perfect, but it was the only animal model on which to test HIV approaches.

That is, until the humanized mouse model came along. Mice have long been a favorite in animal studies because it's easy to obtain large numbers of them and they're not expensive to maintain. The problem with mice, however, is that they're not very much like humans. The strain of mouse most commonly used in research studies, called Black 6, has an 85 percent similarity to the human genome. Compare this to macaque monkeys, which are 95 percent similar. The differences don't lie just in the genes; they also lie in the expression of those genes, many of which, particularly for the immune system, are dissimilar. This was shown in a study looking at inflammatory disease, in which the expression of genes between human and mice didn't match up. These differences had clinical consequences: Of 150 drugs developed in mice to treat sepsis, not one was effective in human trials.

If only there was an animal model that combined the ease and cost of the mouse model with the clinical relevance of the monkey model. As no example existed in nature, researchers made their own: the humanized mouse model. In simple terms, this model takes a mouse genetically altered to have no immune system of its own and transplants human cells or tissues into the animal. Because the mouse has no immune system, it can't reject the human tissue. Instead, the human cells multiply, forming a stable human immune system inside the mouse.

This model is sometimes made using stem cells. The human stem cells home to the mouse bone marrow and take up residence. From here, they form all the cells of the human immune system, creating elaborate tissue networks throughout the blood, kidney, liver, thymus, gut, lymph nodes, and even brain. In some mice, they create a tissue where none existed before. In mice that have no thymus, for example, stem cells can form the organ themselves, a tight cluster of brand-new human tissue that pumps out mature T cells.

As you can imagine, these mice are fragile. Several debilitating mutations give them the ability to be engrafted with human cells. A typical mutation is SCID, more commonly known as the boy-in-the-bubble disease. Babies born with this disease have no functioning immune system and usually die in their first year of life. Those who do survive are kept in sterile conditions. The same rules apply for mice with this disease; they must be kept in sterile conditions and treated gently.

So when Carl June wanted to test the CCR5-cutting ZFNs he got from Sangamo, he decided to use humanized mice.

June and his team took human commander T cells and treated them with Sangamo's CCR5-cutting ZFNs. He then expanded the T cells in culture and transplanted them into a humanized mouse model. When he challenged the mice by infecting them with HIV, he found a new world of natural selection was happening inside the mice. The HIV was killing off the T cells untouched by the ZFNs, while the T cells treated with the ZFNs survived. The efficiency took a giant leap, from only about 10 percent of the CCR5 gene being knocked out, to over 50 percent. Finally, here were reasonable numbers to fight off the virus.

After a month of HIV infection, mice that received cells treated with ZFNs had much lower levels of virus compared to those receiving the control cells. The average viral load in the treated population was 8,300 copies per milliliter, while the controls had 60,100 copies per milliliter. In addition, the CD4+ T cells were substantially higher in those mice that received the unique treatment. All signs pointed to ZFNs as a new gene therapy for HIV.

From his work developing gene therapies, June knew what he needed to do to bring a new drug into clinical trials. But what he found was that the systems for funding cancer and HIV were quite different. It wasn't easy to convince funding and regulatory agencies to take the leap of faith on such an unconventional therapy to treat HIV. At an NIH RAC (Recombinant DNA Advisory Committee) meeting, he remembers the challenge of convincing the group that they could find the right patient population. However, once news broke of the Berlin patient Timothy Brown, attitudes began to change. Suddenly, there was a real-life example of how changing someone's genes could cure HIV. Gene therapy had gotten a shot in the arm from an unlikely source. June calls Timothy's story the "tipping point," saying, "After Tim, you could talk about gene therapy in public. You could get funding for it." The timing couldn't have been better. The news had come just as June was trying to organize the first clinical trial to use CCR5 ZFNs as a therapy for HIV. Of Timothy's story, he says that if "you define a miracle as being a very rare event it was certainly that."

June's CCR5 ZFN clinical trial began in 2009. The Phase I clinical trial tested the experimental therapy in several different treatment groups. The first group, or cohort, was six patients who had failed two different drug regimens. Here was a group of HIV-positive people who were in need; antiviral drugs weren't working for them. Perhaps a gene therapy could swoop in and save the day. This group would receive a single dose of 5–30 billion of their own T cells that had been modified by the ZFNs to no longer express CCR5. Hopefully, once reinfused in the body, those cells would have a selective advantage in the face of HIV and, like Timothy's cells, keep the virus at bay.

The second cohort of patients would be more typical of those infected with HIV. This group of six people would be on normal suppressive therapy. Unlike the first cohort, they were doing fine on antiviral drugs. They, too, would receive 5–30 billion of their own cells engineered to be genetically capable of fighting off HIV. However, this group would experience a treatment interruption, a twelve-week period when they would stop taking their drugs. The idea here was that, for the gene therapy to work, the researchers had to put genetic pressure on the virus. Just as the mice experienced a rise in genetically modified cells only in the face of the virus, the same conditions had to be replicated in humans in order to see any effect. They had to create the right environment for selective pressure, and that meant they needed the virus.

The third cohort of patients was on normal suppressive therapy, but, although the drugs had banished their virus, it had not brought their T cells back to optimal levels. This group of six would be given the same dose of their engineered cells but, unlike the second cohort, they would not be put to the test by stopping their therapy. It was too risky in a group like this.

The 18 patients came to the clinic twice to have their T cells extracted, a harmless procedure, similar to a regular blood draw. They also had a rectal biopsy performed both before and after therapy to assess whether the engineered cells would make their way to their tissues. Five weeks after they first came in to give blood, they would receive the new and improved T cells, infused back into their veins. All the patients would be closely monitored. Four weeks after receiving the infusion, the second cohort would stop taking antiviral drugs for twelve weeks, a treatment interruption designed to give the engineered cells a selective advantage.

This clinical trial, as it was the first to use ZFNs, was primarily designed to assess the safety of the new technology, not to test the efficacy of the gene therapy. For this reason, participants wouldn't be completely stopping their antiviral drugs, the only way to fully test the gene therapy. June and his group found that the ZFNs were safe; the gene-editing machines didn't target any genes they weren't supposed to and didn't cause any adverse reactions. In addition, the engineered cells had made their way to the mucosa in the gut, an important part of any therapy attempting to cure HIV.

When it came to treating HIV, the results mimicked the findings in the humanized mouse model. The cells that were vulnerable to HIV were killed. The cells modified by the ZFNs to be resistant to HIV survived. This was similar to Timothy's case, in which the virus killed off his cells but spared the new ones that didn't express CCR5. It was also similar to the mice, where the presence of HIV was needed to expand the genetically modified cells. When the modified cells expanded, they were able to reduce virus and boost T cells. These results were seen only in the patients who temporarily stopped taking their drugs, the second cohort. This is probably because the gene therapy needs the selective pressure of the virus itself in order to exert control over it. The point is made more apparent in the case of one member of June's clinical trial, who, like Timothy, is hetrozygous for the $\Delta 32$ mutation with one functional copy of CCR5 and one copy that doesn't work. This man, called the Trenton patient, already had a leg up, genetically. He was able to completely control the virus without antiviral therapy. Like Timothy, his CCR5-negative cells controlled HIV. Unlike Timothy, he'd been given this advantage from a gene therapy.

The findings were unprecedented, the first successful gene therapy trial for HIV, prompting June to use the once forbidden

c-word, when he said, "The data obtained in our treatment interruption studies are very exciting and represent significant
progress toward a 'functional cure' for HIV/AIDS." A gene therapy like the one June did is not without risk. However, this type
of therapy presents fewer fiscal and medical challenges than a
bone marrow transplant. The cost of this approach is approximately $300,000 less than the lifetime cost of taking antiviral
drugs. Because it involves manipulation of one's own cells, this
type of procedure can be performed safely in a clinic.

June is excited about the future of this study. The next step is
obvious; June knows they have to take patients off antiviral drugs.
The more engineered cells a patient has, the better he does at
controlling the virus on his own. The only way to get more engineered cells is to lengthen the time the patients are off therapy,
with the possibility of ending therapy altogether. The patients for
this Phase II clinical trial, the first with the real possibility of curing HIV through gene therapy, enrolled in 2013.

June sees his cancer and HIV clinical trials as constantly in-
forming each other; he sees the data from his diverging trials as
a "cross-pollination and fertilization of ideas." Despite this, he
says, cancer and HIV trials are not treated the same; it's far more
difficult to get approval and funding for new HIV therapies.
June says this is the "best and worst time to be in science." It's
the best because of how promising the science is, and the worst
because of the lack of funding to pursue that science.

In this time of shrinking funding for science, June worries
about the investment needed to translate his promising clinical
trial results into a tangible therapy for HIV-positive people everywhere. June has noticed a lack of motivation from big pharma, the

typical powerhouse that brings new therapies to market. At a time when pharmaceutical companies are bringing in considerable profits from their portfolio of antiviral drugs capable of keeping HIV at bay, there is little drive to make the larger investment needed to bring a cure for the disease to the marketplace.

June is optimistic. He believes that "attitudes can change on a dime with a few successful patients" and thinks he has those patients. He hopes that private investors will take the first leap of faith needed to bring the therapy to a wider group of patients. Once this happens, June reasons that the pharmaceutical industry will follow. Sadly, he knows that the NIH, the government source of medical research funding, simply doesn't have the budget to invest in a cure for HIV. The question remains: We can cure HIV, but will anyone fund it?

In 2012, Timothy visited June's lab at the University of Pennsylvania. On a wall of the lab, Timothy's picture is displayed. June says it's "almost like a religious thing." He sees Timothy as "raised from the dead." As Timothy moved about the maze of lab benches and tissue culture rooms, his very presence seemed to inspire the students and technicians around him. In June's words, "An $n=1$ is an incredible thing." In science, $n=x$, means the number (n) of participants (x) in a study; in a field driven by data, the higher the number of participants, the more convincing the results. But Timothy's case is the exception. The power of his story rises above statistical significance. Scientists are only human. Sometimes the sway of a great story is as powerful as the most comprehensive dataset. We can't underestimate the impact a story has on the course of science.

CHAPTER 26

A Child Cured—So What?

I n early 2013, researchers were stunned to open their program book to paper #48LB at the 20th Conference on Retroviruses and Opportunistic Infections, held in Atlanta, Georgia. The paper was a late-breaking abstract, meaning that the data was brand-new, never published. The abstract, entitled "Functional HIV Cure after Very Early ART of an Infected Infant," was an exciting development. The abstract's opening line was: "A single case of HIV cure occurred in an infected adult with a bone marrow transplant," referring, of course, to Timothy Brown. The case being reported was radically different from Timothy's. The child had, only 30 hours after being born, received three antiretroviral drugs: Jerome Horwitz's AZT, 3TC (lamivudine), and neviraprine. On the second day of life, doctors detected HIV in the baby girl, and tested on a weekly basis thereafter. On four successive blood draws, HIV was measured using a sensitive PCR-based assay. The baby was HIV-positive. But, to the surprise of her physicians, the virus slowly disappeared in the baby, becoming undetectable by day 30. Now, two years later, doctors felt confident in calling the child cured.

There was that *c*-word again. But now Timothy's influence was so great in the HIV community that physicians were no longer afraid to use the word *cure*. Unacknowledged but just as influential was Christian, the first Berlin patient, the impetus behind the first clinical trials testing early therapy. Here in this one child lay the combined promise of both Berlin patients. The eradication of the viral reservoir was an experience shared by both Christian and Timothy. The baby had received a very early therapy, right after being infected by the mother, similar to the early therapy Christian received. Because of such an early therapy, the reservoir didn't have a chance to take hold in the child. However, similar to both Berlin patients, a small amount of virus was detected by ultrasensitive PCR in a subset of blood cells, called monocytes, taken from the baby. Here, then, was the unifying lesson taken from both Berlin patients: We don't need to eliminate the virus completely. We can create a functional cure for HIV by eliminating enough of the virus, whether it is by early aggressive therapy like Christian received or a gene therapy inspired by Timothy.

Anthony Fauci, director of NIAID, believes that early therapy, such as this baby received, is the path to a cure. Speaking about the promise of this kind of therapy, he says, "Children will be the first group to be cured." This is a powerful statement coming from Fauci, who is notoriously guarded about making such claims. Fauci says he's a scientist, and therefore, "I don't jump out of my pants for anything." Despite his scientific skepticism, an important quality for any person who holds such influence among researchers and policy makers alike, Fauci says, speaking of

Timothy, "Just having a person who is cured galvanized a lot of enthusiasm." Once again, sometimes it's the influence of the story that means more than the data.

In April 2013, a twelve-year-old boy at the University of Minnesota Medical Center, Eric Blue, received an infusion of cells that looked an awful lot like Timothy's. These cells are hematopoietic stem cells like those Timothy received, but this time, instead of coming from a stranger's bone marrow, they came from a baby born into the world. These stem cells originated from cord blood, the blood collected from the umbilical cord and placenta when a baby is born.

Hematopoietic stem cells, the progenitors that form all the cells of our immune system, are found at high concentrations in cord blood, ten times higher than in bone marrow. Better still, unlike obtaining stem cells from bone marrow, which requires a surgical procedure, cord blood stem cells are found in the discarded by-products of childbirth. And, while bone marrow cells must be matched exactly between donor and recipient, cord blood cells don't have to be, because the cells are more primitive than those extracted from adult bone marrow. The primitive nature of cord blood cells also means that the procedure is less risky than bone marrow transplants; patients receiving cord blood transplants are less likely to suffer from graft-versus-host disease, the deadly disease in which transplanted cells attack the host's body.

Scientists estimate that the chance of finding a matching bone marrow donor with the $\Delta32$ mutation is 1 in 10 million. Because cord blood doesn't have to be perfectly matched to a patient's blood, it's far simpler to find a donor with the Δ32 mutation. This is exactly what researchers did. The young boy received chemotherapy and radiation in an attempt to destroy his cancer and HIV-riddled immune system. They then infused the cord blood cells naturally resistant to HIV. Researchers hoped that this boy would become like Timothy, cured of both his cancer and HIV in one fell swoop. An hour after the boy received the mutant cord blood cells, Timothy gave him a call to wish him luck and give him some advice: "Make sure as soon as you are able, get out of bed and do some exercise, go do what you love, go play some basketball."

Lead researcher on the study John Wagner, put the procedure in perspective, saying, "There are patients with HIV and leukemia out there today who are waiting for such a breakthrough. But for those with HIV alone, a success in this patient would compel the scientific community to find potentially safer strategies, such as genetically inducing the variant in the patients' own marrow stem cells."

Sadly, on July 5, two and a half months after his transplant, Eric passed away. The boy developed graft-versus-host disease, the same disease that came close to taking Timothy's life. While cord blood transplants are less likely to develop graft-versus-host, any bone marrow transplant is risky. Eric's death underlines the fact that transplants such as these should only be performed in those who require it for their cancer. His case, despite its tragic ending, has inspired physicians in other parts of the globe to try HIV-resistant cord blood transplants in those patients suffering from both cancer and HIV. However, if we're going to cure HIV in those without cancer, we have to find safer ways to translate Timothy's success.

. . .

After Christian's case received mass attention in the late 1990s, and the subsequent clinical trials based on his therapy failed, the field became wary of the early therapy approach to treating HIV. Despite this warranted skepticism, one researcher remained committed to finding a cure: David Margolis. Margolis graduated from Tufts University School of Medicine in 1985, then stayed at Tufts for his residency. At a time when HIV was taking over hospitals across the Charles River in Boston, Margolis felt insulated from the epidemic at Tufts. He was eager to treat HIV-positive patients; they just didn't have any. That enthusiasm led him to a fellowship in infectious disease at the NIH. There, he was thrown into the fire of HIV medicine. It was a field like no other, giving the young physician the opportunity of applying molecular biology to clinical medicine.

Despite the foibles of those early therapy clinical trials, none of which were able to replicate Christian's unique experience, Margolis was unique in the HIV community for pursuing the eradication of HIV. At a time when it wasn't popular to talk openly about a functional cure for HIV, Margolis was persistent.

Similar to how Jessen pursued an experimental cancer drug to eradicate the virus before the reservoir could take hold, Margolis was pursuing a parallel strategy. He was interested in a group of cancer drugs called histone deacetylase inhibitors. These drugs work by altering our control over our own DNA. DNA is wrapped tightly around proteins called histones. Because our DNA is in such long, unwieldy chains, we need to wrap them around these histones to keep them organized. Similar to how a garden hose is wrapped around a holster in the garden, histone deacetylase keeps the DNA wound around the histones. It's the presence of this enzyme that lets us unwrap the DNA so that information from the gene can be copied and transcribed by the cell. This enzyme is key to how we use our genes.

Cancer researchers postulated that by inhibiting this enzyme, they would fire up the tumor-suppression gene, a gene that, just as it sounds, protects the cell from cancer. This hypothesis was correct: vorinostat, developed by Merck, was the first such inhibitor to be approved by the FDA for use in cancer in 2006.

HIV researchers such as Margolis took note of this use of this kind of inhibitor. Margolis had been working with this class of inhibitors since 1996 when he first discovered how the anticancer drugs interact with latent HIV. As we have seen, the challenge with eradicating HIV from the body is that the virus can hide in our DNA. When it does this, it's called a latent virus because, although it isn't easily detectable, it provides a continual source of virus, essentially a reservoir that standard antiviral drugs aren't able to shrink. Even after a person has spent decades on antiviral drugs with no detectable virus, once they stop taking the drugs, the virus comes back. Thanks to the viral reservoir. Margolis believed that histone deacetylase inhibitors had the potential to root out the virus hiding in our DNA, just as it woke up the tumor-suppressor gene in cancer. By unwinding the DNA, the drug could uncover the virus hiding within. It was the early 2000s and vorinostat was not yet available. The only inhibitor like it that was available was valproic acid, a drug used to treat seizures and mood disorders.

In 2004, Margolis enrolled a pilot study of four patients who consented to taking the unusual therapy twice a day for three months. Margolis and his group then quantified the amount of HIV in the resting T cells. These immune cells that aren't actively dividing represent the biggest challenge to eradicating the virus. If Margolis could get them to release HIV, he knew he was on to something. Three of the four participants had a sizable reduction in the latent viral reservoir, on average 75 percent. Margolis and his colleagues made a splash in the HIV community and in the popular press when they published these results in *The Lancet* in 2005. But the excitement was short-lived; the promising reduction in the viral reservoir faded. Eight months after first receiving valproic acid, the reservoir was back. Like so many eradication attempts before it, valproic acid worked better in theory than it did in patients.

Many scientists faced with such disappointing data would have given up on histone deacetylase inhibitors, but not Margolis. The problem, he reasoned, was finding the right drug. He turned his attention to a different inhibitor known to have a potent effect on multiple classes of the enzyme: vorinostat. It was a drug he had long been interested in but was only recently available to test in humans. Unfortunately, vorinostat isn't as benign as valproic acid. Vorinostat had the ability to cause mutations in DNA, which can possibly lead to cancer. Getting the FDA to allow Margolis to test the drug would require three years of convincing.

In a packed conference room at the 19th Conference on Retroviruses and Opportunistic Infections, in Seattle, Washington, in 2012, attendees eagerly awaited the results of Margolis's vorinostat trial. The room was characterless, like any conference room, anywhere in the world. But inside, the crowd was electrified. They knew what was coming. For months, HIV researchers had been gossiping about Margolis's trial and its promising results. Now the results were to be announced in the late-breaking abstract. The conference room couldn't hold the throng anxious to hear the results; attendees spilled out into two overflow rooms.

Margolis's vorinostat results were small but impressive. Margolis had 6 participants. Those 6 HIV-positive men had been given only a single dose of the drug, for the FDA restricted its use. Given these constraints, the community wasn't expecting what followed. In all 6 men, the reservoir of virus in the resting commander T cells increased, on average fivefold. It was a sign that the drug was releasing the latent virus hiding in the T cells. In a press release put out by the university at the time, Margolis said, "This proves for the first time that there are ways to specifically treat viral latency, the first step toward curing HIV infection." His results were echoed by those of Sharon Lewin, a researcher from Australia, who found similar safety and efficacy results in her small trial of vorinostat. Researchers hope that the promise of Christian's cure can be fulfilled in this new but similar eradication strategy.

Margolis, who now has positive results in 8 patients, has convinced the FDA to allow him to test vorinostat on a larger scale. In an ongoing clinical trial, study participants will be given the drug three times a week for eight weeks.

"Cancer, diabetes, multiple sclerosis, you can have any of these diseases but they don't make you the 'other' in the way that HIV does." Margolis's patients still ask him, "When is there going to be a cure?" It is as if no one had ever been cured of HIV.

CHAPTER 27

Zinc Finger Snap

Both Heiko Jessen and Gero Hütter shared an interesting characteristic when they came up with their unique therapies that the Berlin patients received: They had little experience. Hütter had never treated an HIV patient, while Jessen was testing out the early aggressive therapy informally, with no previous experience in conducting a clinical trial. Similarly, when Sangamo gave Paula Cannon, a researcher at the University of Southern California, their CCR5 ZFNs to test, they couldn't have been expecting much.

Cannon had no experience in gene therapy or animal models. Lack of experience had never stopped Cannon yet. She'd worked as a rock band manager and wedding dressmaker before becoming a scientist. With her charming English accent, quick wit, and catalog of A-list collaborators, Cannon was able to convince the small biotech company to let her give the CCR5 ZFNs a try. Her pitch was ambitious. She proposed using the ZFNs in hematopoietic stem cells, the stem cells that give rise to all the immune cells in the body. Those stem cells would then be engrafted in a humanized mouse model and challenged with HIV.

It was quite a proposal, considering that Cannon had never worked with stem cells or a humanized mouse model. She was a young assistant professor who had a small lab and a modest budget. Still, despite these drawbacks, Sangamo sent her, as well as other collaborators, the CCR5 ZFNs. There was little risk for the company, Cannon would either deliver the data or not. If she didn't, then it was likely that a different collaborator would. Bringing an inexperienced graduate student, me, on, and with a tiny $50,000 grant from the California HIV/AIDS Research Program, Cannon was able to do what larger, better-funded labs could not: treat the finicky stem cells with the ZFNs, engraft them in humanized mice, and challenge with HIV. The virus put extreme pressure on the immune system. The results were phenomenal: Mice given the ZFN-modified stem cells developed a human immune system devoid of CCR5, without which the virus couldn't enter the T cells. The mice receiving the gene therapy cleared their HIV infection. In contrast, mice that received mock-treated cells, that is, cells that received all the same manipulation but without the CCR5 ZFNs, had high levels of HIV and progressed to AIDS.

The convincing results were published in *Nature Biotechnology* in 2010. All Cannon needed were the right clinical collaborators to bring the technology to human trials. That's when she was introduced to John Zaia at City of Hope hospital in Duarte, California. Together they formed what the CEO of Sangamo would call "the dream team." The two came up with a bold plan to translate the dramatic findings from the humanized mice into humans. They postulated that the best population to test such a therapy was in HIV-positive patients who, like Timothy, had AML (acute myeloid leukemia). These were patients who needed a stem cell transplant. They couldn't find donors who had the Δ32 mutation, so they would do the next best thing: make the stem cells look as if they came from a person naturally resistant to HIV. They would then infuse the cells back into the patient. The stem cells would travel to the bone marrow, where they would form all the cells that make up the human immune system. Like Timothy's experience, and Carl June's promising data, the group believed, the cells would have a survival advantage when faced with the virus. Inspired by Timothy, the group believed they could create a functional cure for HIV.

It was a bold plan and an expensive one. The safety studies needed before they could bring the new technology to a clinical trial were not trivial. Then the trial itself would be costly. This was a problem because, while the NIH funded basic research, they shied away from supporting advanced studies heading to a clinical trial. The group applied for a new kind of funding made available by CIRM (California Institute for Regenerative Medicine). Governor Arnold Schwarzenegger had formed this institute in reaction to George Bush's freeze on federal funding for stem cell research. The data generated from Paula Cannon's tiny $50,000 grant garnered the top score at CIRM that year. The grant, which referenced Timothy's case, brought the funding the project needed: a whopping $14.5 million.

Cannon still finds it funny that anyone is surprised by her results. She says, "The fact that it worked, it was like the 'Duh!' moment. It was the most unremarkable thing. I wasn't quite prepared for how exciting people thought the results were."

. . .

In 2008, Timothy Henrich was a busy second-year resident in internal medicine at Brigham and Women's Hospital in Boston when he first heard about the Berlin patient. He immediately knew that this was where HIV research was heading. He thought the Berlin patient was "the most exciting development since antiretroviral therapy." A young doctor interested in infectious disease, he wanted to be a part of the biggest thing to hit HIV in a decade, when the Berlin case was changing how HIV researchers saw the future, and the word *cure* was coming back into favor. Unfortunately, Tim was busy, his schedule packed with the heavy demands of his residency, leaving little time for research.

Two years later, Henrich was doing his fellowship in infectious disease at Brigham and Women's Hospital when he began looking for a research project. His interest in the Berlin patient had only grown in the intervening years, and so had his need for a successful project. Henrich was in the precarious position of being a young researcher, with limited funding, desperate for a project that could bring him the publications and grants he would need to become faculty at the hospital. It's a stressful time for any scientist starting out in their field. Funds are limited; time is precious; there are simply not enough faculty positions to go around. Under such pressure, many physician scientists tend to take the easy route, performing science that can be done quickly, to get as many publications as they can.

Henrich knew he needed a project that would get him published and bring in grants, but he didn't want to compromise on the science. After a failed project attempt, he decided to follow in the footsteps of Gero Hütter and the Berlin patient. If scientists wanted to turn Timothy's treatment into one accessible to HIV-positive people everywhere, they needed to understand what role each component of his treatment played in Timothy's ultimate cure. Timothy had received chemotherapy, a conditioning regimen, and a bone marrow transplant; had graft-versus-host disease; and received donor stem cells with mutant CCR5. All signs pointed to the mutant CCR5 as the cause of Timothy's cure. This is because Timothy went from having a Δ32 mutation in one copy of the mutant gene to having the mutation in both copies. It made logical sense that the change in Timothy's genotype corresponded to a selective pressure exerted by the virus itself and was ultimately responsible for his ability to clear the virus. Although this made perfect sense, no one could be sure that the other factors involved in his therapy hadn't influenced the outcome. Could the intense conditioning regimen have cleared out the virus? Or could the bone marrow transplant itself have caused the dramatic effect? This was Tim Henrich's question.

With his advisor, Dan Kuritzkes, the director of AIDS research at Brigham and Women's Hospital, they began to look for patients who would fit the bill. They needed to find people who had HIV but also, as a medical necessity, needed a bone marrow transplant. They wouldn't try to find a donor who was naturally resistant to HIV, as Timothy was. Their goal wasn't to cure HIV. Instead, they wanted to see what effect getting a bone marrow transplant had on the HIV reservoir. They hypothesized that the transplant itself, as it swapped out so many of a person's own immune cells, would perturb the reservoir. It's similar to

Hütter's notion of "resetting the immune system clock." By doing so, they might also determine which cells were key to maintaining the virus reservoir.

Henrich's study started as retrospective but, after the spectacular results, became prospective. He was working with archived samples from patients who'd already had the procedure. Quite by accident, the researchers found archived samples from two HIV-positive men with lymphoma. The men had received minimal ablation, the drugs given to clear out a person's own cells in their bone marrow to make way for the transplant. This was in contrast to Timothy, who received an aggressive ablative conditioning regimen. Because the ablative treatment was minimal, the men were able to continue taking their antiviral drugs. Timothy's more aggressive treatment and chemotherapy meant that he had to stop therapy. However, similar to Timothy's experience, the donor cells homed to the patients' bone marrow and, over time, replaced the men's own immune cells.

What Henrich's team found was unexpected. They had hoped to model the decay of the HIV reservoir in resting commander T cells, the cells that unwittingly hide the virus in our DNA, out of reach of current antiviral drugs. What they found, however, was no latent virus at all. The two men, who received the therapy 2.5 and 3 years ago respectively, appeared to have eradicated their viral reservoir. It was an exciting announcement at the AIDS conference in DC in July 2012. It seemed that the promises made by one of the Berlin patients had finally been fulfilled. The word *cure* was passed around as the story grabbed headlines. NPR ran the story under the headline TWO MORE NEARING AIDS

CURE AFTER BONE MARROW TRANSPLANTS. But in actuality, the story was still more complicated.

The two men had not stopped antiviral therapy, it was unknown whether the virus would rebound. Also, while Timothy had biopsies performed in his brain, gut, and lymph nodes to search for the HIV reservoir, these two Boston patients hadn't yet had any additional biopsies done. This is an important point, for HIV is known to hide in these T-cell-rich tissues.

Even if these hurdles were overcome, there are other reasons why this approach couldn't be used for the majority of people living with HIV. The main drawback, as we've discussed, is the considerable risk involved in bone marrow transplants. As Henrich himself says, "If you don't need a bone marrow transplant you shouldn't get a bone marrow transplant."

What this study makes clear is a feasible path to eradicating the virus. These researchers were able to shrink the viral reservoir, the obstacle to curing HIV. While bone marrow transplants will never be commonly used to eradicate HIV, this approach leads the way for other technologies, such as gene therapy and histone deacetylase inhibitors.

Asier Sáez-Cirión, an assistant professor at the Pasteur Institute in Paris, was discontent with the studies on the benefits of early therapy in HIV. He wasn't alone; many in the field have expressed frustration at the fact that we still can't recommend early therapy and are unsure what benefits it may give, if any. To address this, Sáez-Cirión decided to go back and look at the records of 700 French HIV patients who received early therapy. With Christian's story as inspiration, the patients in the late 1990s were given antiviral therapy during acute HIV infection. Sáez-Cirión's is a retrospective study, requiring no new patients. The advantage to this kind of study is that you can look at large numbers of patients with little cost; the disadvantage is that you can't modify the study since it's already happened. Of the 700 patients, 75 of them stopped therapy after a year. Of those 75, 14 of them had not returned to therapy. These 14 became known as the VISCONTI cohort (acronym for virological and immunological studies in controllers after treatment interruption).

These 14 patients have a few unique features. They all started therapy very early: The median time to start therapy after infection was 39 days. While not as early as Christian, who received therapy within days of his infection, the cohort was treated earlier than other studies on acute HIV at the time. The VISCONTI cohort stayed on therapy between one and seven years before stopping. This, too, was different from other trials that treated for shorter periods, similar to Christian's self-prescribed treatment interruption, which came only six months after starting therapy. Unlike Christian, the therapy the cohort received was standard and included no experimental cancer drugs. Like Christian, many of the patients experienced a short-lived spike in HIV after stopping therapy. Unlike Christian and elite controllers, the storm trooper T cells from these patients didn't have any special ability to target HIV.

Approximately seven years after the cohort stopped therapy, researchers announced the results in 2012 at the AIDS conference in Washington, DC. The 14 patients remained off therapy.

Because none of the patients had any genetic markers that could explain control of the virus, they were declared, like Timothy and Christian, functionally cured. Intriguingly, they also harbored a tiny amount of virus detectable by ultrasensitive PCR in their T cells, just like Timothy and Christian. Perhaps even more surprising was that in four of the patients, this tiny pool of virus continued to shrink, even though the men hadn't been on therapy in years.

The pieces of the puzzle were coming together. The evidence from the VISCONTI cohort tied in perfectly with the anecdotal evidence offered from the Berlin patients as well as the toddler who was given early therapy that resulted in a functional cure. It also matched the data from the trials of histone deacetylase inhibitors. The answer wasn't a sterilizing cure for HIV infection; not every bit of virus has to be eradicated. Instead, it's possible to live with some HIV still hiding in the body, a tiny pool of passenger virus that was along for the ride but needed no special effort to be contained. The path to a functional cure for HIV was taking many forms, from gene therapy based on Timothy's experience, to early therapy based on Christian's. But it's all leading to the same place.

David Baltimore, to put it mildly, has been interested in retroviruses for a long time. In 1975, he was awarded the Nobel Prize for discovering reverse transcriptase, work he did during his postdoctoral fellowship that revealed how retroviruses invade our DNA. Even back then, he saw the potential, reminiscing, "One of the things that struck me immediately when we found

reverse transcriptase is that it was a door to gene therapy."
Those who did pursue gene therapy during those early days ran
into difficulties, for the field was simply too new. But the promise was there. Researchers had figured out how retroviruses gain
entry to cells and insert their genetic material into our DNA.
Perhaps there was a way we could manipulate this system to
insert a gene of our choosing into our DNA.

Baltimore was performing basic immunology research when
once again he was intrigued by the promise of gene therapy. In
the early 2000s, he partnered with Irvin Chen at UCLA. Together they tested the ability of small interfering RNAs (siRNAs)
to inhibit CCR5. These short RNA molecules are able to inhibit
gene expression through a mechanism known as RNA interfer-
ence (RNAi). The small pieces of RNA bind to the messenger
RNA (mRNA), which keeps the message, the gene information
needed, from reaching the ribosome, the protein construction
plant. The mRNA is like a message in a bottle, a necessary mis-
sive the cell needs to express its genes. The siRNA breaks that
bottle so that the message is never delivered. Without the CCR5
message delivered, the protein can't be expressed on the surface
of the cell. This means that HIV can't enter the cells—just as a
person with a mutant CCR5 doesn't express the protein on the
surface of his T cells. Baltimore and Chen's results, published in
2003, were promising. But the research stayed on the shelf because the next step was a human clinical trial, an expensive un-
dertaking. "It wasn't clear we could get support for it," says
Baltimore. Years later, he met an entrepreneur, Louis Breton,
who was interested in the approach. Together they formed a
small biotech named Calimmune in 2007. But they still needed

funding to bring it to clinical trials. This wasn't so easy since a gene therapy approach to treating HIV was generally considered a risky investment.

That changed when news of the Berlin patient, Timothy, broke in 2009. Suddenly, gene therapy approaches no longer seemed outlandish. This influenced not only researchers but organizations that fund researchers, such as amFAR, The Foundation for AIDS Research. In fact, when discussing the grant they awarded Baltimore in 2010, they state, "amfAR's interest in exploring the role of gene therapy in the eradication of HIV infection stems from a February 2009 report in *The New England Journal of Medicine* of a patient in Berlin." But the grant that changed the fate for Baltimore's CCR5 siRNA approach came from CIRM, the California State–funded stem cell agency. The $20 million grant was awarded in 2010 to bring their encouraging gene therapy to clinical trials. CIRM awarded the grant, and the one for Cannon and Sangamo, based on the hopes bolstered by Timothy. The project began enrolling the first patients in March 2013. Now multiple gene therapy clinical trials, all targeting CCR5, are in motion, all based on Timothy's cure.

CHAPTER 28

The Abused, the Respected, the Relentless

G ero Hütter's paper begins: "A 40-year-old white man with newly diagnosed acute myeloid leukemia (FAB M4 subtype, with normal cytogenetic features) presented to our hospital." Yet behind the cold scientific facts lay a mosaic of human experience. Timothy's journey didn't stop when he was cured of HIV. He wasn't the same person as when he first entered Charité hospital in 1995. He had gone through chemotherapy, a brain biopsy, ablative therapy, and a bone marrow transplant. It was enough to change anyone. Timothy walks with a slight limp and speaks slowly and softly. He occasionally gets confused, a side effect that will lessen as the years go by.

In 2011, Timothy moved to San Francisco. He was excited to return to his native country after spending the last decade in Germany. But things weren't as easy in the States as they were in Europe. In Germany, Timothy received money from the government for his housing, food, and medical care. Everything was covered. This was important because, given Timothy's condition, he couldn't work and he still needed significant medical care.

Many people believe that because Timothy Brown is the Berlin patient, widely touted as the first person cured of HIV, he must make a decent living. This couldn't be further from the truth. Timothy lives in a run-down government apartment building outside Chinatown. His neighborhood is not safe nor is his apartment building, whose residents frequently have problems with violence and drug abuse. His apartment is a single cramped room with just enough space for a twin bed and a hot plate. Unpacked boxes line the walls; there is no place for Timothy's meager belongings. It's difficult to keep pests out; his mattress is full of bed bugs. He has a tiny attached bathroom and access to a larger communal kitchen down the hall, which is too disgusting to discuss, much less cook in. In some ways, the apartment is barely his own; he is limited to the number of nights a week his guests can spend with him, limiting the amount of time he has with his boyfriend.

Timothy recently spoke with Marcus, the man who told him he needed to get tested for HIV eighteen years ago. As Timothy shared his story, how he was cured of HIV, he could sense Marcus recoil. "But who cares about curing HIV?" Marcus asked. Marcus had been on antiviral drugs for over a decade and couldn't contemplate the millions of HIV-positive people who have trouble taking drugs or don't have access to them. "You're wasting your time," he said to Timothy. Timothy was hurt by these words. Timothy's hope was that by sharing his experience, an experience that was difficult, he would create enthusiasm for the long-awaited cure for HIV. To have a friend tell him that a cure wasn't important wounded him.

Timothy is generous with his time, speaking at conferences

in both the United States and Europe, usually without compensation. The audience has little idea that the man before them is barely able to pay his bills. Timothy is also generous with the donation of his blood and tissues, which he gives regularly to the lab of Steve Deeks at the University of California, San Francisco. Scientists regularly test his blood and rectal biopsy samples by ultrasensitive PCR for traces of the virus. Given that Timothy had traces of a CXCR4-utilizing virus, researchers believed the virus would rebound quickly in him. That's because the donor cells Timothy received were naturally resistant to CCR5-utilizing viruses, not to CXCR4. All along, researchers had been warning Hütter that this would happen because CXCR4-utilizing viruses tend to pop up late in infection and cause rapid disease. The presence of CXCR4-utilizing virus in Timothy's gut was a sure sign that the virus would grow and Timothy would have to restart his antiviral drugs.

Surprisingly, this did not happen. And no one knows why. Researchers postulate that perhaps CXCR4 viruses need some modification or weakening of the immune system from CCR5 viruses in order to grow. Although this doesn't explain the rare cases in which individuals have been infected with CXCR4 and not CCR5. Some have postulated that perhaps the $\Delta 32$ mutation confers some resistance to CXCR4 viruses as well, altering some trafficking of the chemokine that we don't understand. Probably the most logical answer is that there is some level of virus we can control. Although it's difficult to quantify what the tipping point is, there's a measure of virus we can live with without ill effect. This goes hand in hand with the experience of Christian, who also has a tiny amount of detectable virus in his resting T cells and lymph nodes. Despite this, he hasn't had to take medication in fifteen years. Similar as well is the toddler who was declared functionally cured but has some detectable HIV in her resting T cells. Again, this is the real point. We may not be able to eliminate every trace of virus in a person, but we don't have to. We simply need the right tools, be they gene therapy inspired by Timothy or early antiviral and eradication therapies inspired by Christian, to bring the virus down to a level we can deal with.

Steve Yukl, a colleague of Steve Deeks's who works closely with him on Timothy's case, made this fundamental point in 2012 at a small workshop on HIV in Sitges, Spain. Yukl had just announced some unusual results. He had sent out Timothy's samples to multiple collaborators spread across the country to test for HIV. Using a highly sensitive PCR test for HIV RNA, they detected a low signal. The results, he cautioned, weren't consistent, and because of how the assay was done, they weren't reliable. In fact, he added at the conference, it was likely they had been contaminated. PCR takes advantage of the natural ability of DNA to bind and the power of the polymerase enzyme in order to produce infinite copies of a specific gene or other target. While PCR can be highly reliable, the more times it's repeated on a single sample, the more unreliable it can get. This is because, by repeating the reaction multiple times, you use less and less of the original sample.

Timothy underwent numerous procedures and biopsies to give these samples to science. The samples are taken from his blood, rectum, ileum, lymph nodes. He even underwent a lumbar puncture to get a sample of his cerebrospinal fluid. The number of cells taken from each of these procedures is small, so the

RNA amplified from them needs to be amplified an unusually high number of times. The more cycles of PCR done, the more likely a false positive. In an interview with *Science*, Douglas Richman, an HIV researcher at the University of California, San Diego, explained it this way: "If you do enough cycles of PCR, you can get a signal in water for pink elephants."

Other problems in the analysis occurred. When different collaborators sequenced the virus they had amplified by PCR, it didn't match the original virus Timothy was infected with. However, it also didn't match among the collaborators. This was a sign of contamination. It was obvious that the results needed to be repeated since they brought up more questions than answers. Steve Yukl decided to share the preliminary results from Timothy's samples as a means to discuss HIV reservoir with the group. Perhaps naively, he didn't expect that the small group of scientists would be misled by the data. Why was this a big deal? Any sign that the Berlin patient might not actually be cured was sure to grab headlines. Although for those who were familiar with the case, it wasn't news that Timothy might have virus hidden in his body. After all, in the original *New England Journal of Medicine* paper, Hütter discussed the traces of CXCR4-utilizing virus hidden in his gut. Technically, this wasn't even news.

On June 11, 2012, a French HIV researcher who attended the meeting in Spain issued a press release with the headline THE SO CALLED HIV CURED "BERLIN" PATIENT STILL HAS DETECTABLE HIV IN HIS BODY. Contrast this headline with the title of Yukl's talk at the meeting: "Challenges inherent in detecting HIV persistence during potentially curative interventions." The press release mentions none of the caveats that Yukl made when presenting

the now infamous data. It does not mention the strong possibility of contamination. Instead, it presents Yukl's results as a
"challenge" to Hütter's cure data. Both Yukl and Deeks were
perturbed when they saw how their data had been twisted in
the media. In an interview with *Science*, Yukl sought to clear up
the controversy, saying, "The point of the presentation was to
raise the question of how do we define a cure and, at this level
of detection, how do we know the signal is real?"

The press release reads, "These data also raise the possibility
that the patient has been reinfected." For Timothy, and others
reading it, this suggestion was an insinuation about Timothy's sex
life, since the only logical way that Timothy could be reinfected
is by having unsafe sex. The personal implication of the remarks
made about Timothy reveals an inherent flaw in the relationship
between scientist and research subject. Because our studies keep
a distance between researcher and study subject, we lose our
empathy. For Timothy the effect was humiliating. He watched as
the popular press discussed his sex life and questioned his cure.
Many HIV-positive individuals were also affected. Now confusion swirled around what these results meant and whether the
Berlin patient was truly cured. It's the kind of news that deterio-
rates hope in HIV-positive people who have suffered so many
previous disappointments. It's also the kind of news that shakes
the public trust in science. Since that time, new data has shown
that these preliminary results were false. In fact, when the tests
were repeated no lab could find detectable virus. Yukl has said
that Timothy's treatment surpasses that of Bruce Walker's HIV
controllers: "Even the most extraordinary 'elite' controllers

described in the literature have more robust evidence for persistent infection." He has even gone a step further in saying that Timothy, while he is described as having a functional cure—that is, detectable virus in his body—"may even have had a sterilizing cure." That is, no remaining virus at all.

This is not to say that scientists shouldn't question the results they hear at conferences or that they shouldn't openly discuss all the implications of new research. It's important for researchers to do this, for it makes the community stronger. However, we need to consider the human factor when discussing research studies. Timothy is not simply the Berlin patient. In light of all he's given research, he deserves our human respect.

Christian is in a very different place in his life compared to Timothy. Where Timothy's life is tumultuous, Christian's is stable. He says his life is relatively untouched by HIV. It's the life dreamt of by many HIV-positive people waiting for a cure. Today, he has all he could have asked for at twenty-seven, when he was first infected with HIV. He hasn't taken antiviral drugs in fifteen years. He has a job he loves. He travels the world on exotic vacations. He has a long-term partner whom he cares for deeply. Yet, his identity is muddled.

He considers himself HIV-positive, although he hasn't harbored the virus in over a decade. He is not alone in this identity. Timothy, too, identifies himself as HIV-positive, although he's cured of the disease. It's almost as if the virus carries an identity of its own, and all those who have carried it, no matter how briefly, will have their lives forever changed by it. Being HIV-positive has become part of the Berlin patients' characters, less a disease than a force that defines who they are.

Christian may identify himself as HIV-positive, but he has a harder time identifying himself as the Berlin patient. Given his mild personality, he dislikes connecting himself with the dramatic medical cure. For this reason his long-term partner, Greg, didn't know he was the original Berlin patient until a year after their relationship started. Greg smiles when he describes the moment Christian first asked him to go with him to see Jessen. Greg was nervous; after all, what could Christian have to say to him that needed to be told at a doctor's office? Was he sick? Imagine Greg's surprise when he learned that Christian wasn't sick or contagious. He was the original Berlin patient. Greg remembered the news coverage of the Berlin patient, the fantastic case of a man cured of HIV in Germany. He never would have expected that his boyfriend was at the center of such a dramatic medical story.

Christian and Greg have shared their lives for eight years. They spend holidays with each other's families. They take wonderful vacations. They are the epitome of a happy couple, supported by their loving families. Christian remains in perfect health, having no lasting effects from his HIV infection. He thinks little about HIV research these days and doesn't follow developments in the field. But hidden away in a drawer of his house lies a complex, handwritten schedule from 1996. It serves as a memento of the therapy he endured.

Timothy's life is almost the reverse of this serene picture. His living situation is dreadful. His love life is tumultuous. His health is fragile as he suffers physically from the cumulative effects of his cancer and HIV therapy. Unlike Christian, Timothy is disabled and can't work. Timothy is also committed to being part of bringing his HIV cure to others. In 2012, with support from the World AIDS Institute, Timothy launched the Timothy Ray Brown Foundation, a nonprofit dedicated to raising money for HIV cure research. It might seem an unusual move for a man with no money of his own. Timothy hopes that simply the power of his name and his story will bring attention to risky cure research in an era with shrinking funding for science.

Bruce Walker has found a way to combat the dwindling money for science. He has sought out private investors, angels in disguise, willing to put their money into risky research projects. This private funding, from donors like Mark and Lisa Schwartz, Terry and Susan Ragon, and Bill and Melinda Gates, is filling in the gaps. Without these sources, it's hard to say what would happen to research projects with great promise but limited data. With this funding, Walker has built an entire institute on the shoulders (or rather the blood) of elite controllers. In the pipeline are new therapies and vaccines based on the personal genetics of those whose bodies control HIV.

Gero Hütter's life has also changed in the years since he published his results on Timothy Brown. First ignored, then hyped, then adapted, Hütter's research went on a roller coaster of influence. Given the scholarly articles published and the attention in the press, what happened next is surprising. Charité hospital shut down its transplant program. With funding problems shaking public hospitals and governments all over Europe, the Berlin hospital wasn't immune to the budget cuts. The successful program, the first of its kind to cure a person of HIV, was slashed. While all those in the medical community assumed that Hütter would continue his work and find another HIV-positive person who needed a bone marrow transplant, he was, in fact, looking for a new job.

Today, Hütter is head of the Institute of Transfusion Medicine and Immunology at Heidelberg University in Mannheim. He has identified two other men with cases like Timothy's who are HIV-positive but, because of their cancer, need a bone marrow transplant. He plans to use donors with the mutant $\Delta 32$ version of CCR5 in an effort to repeat Timothy's success. He works with collaborators all over the world, including Sangamo. Although he's the doctor who cured a man of HIV, his salary is modest. When he visits Berlin, he stays in a hostel. Hütter is married and has a son, born in the summer of 2012.

Heiko Jessen still runs his bustling practice in Berlin, working long hours and holidays. He loves his patients and treats them tenderly. He calls his young male patients his baby boys. He worries about them and takes their personal turmoil to heart. He sees his patients as part of his family. His own family is precious to him. He is close to both his brother, a fellow physician in the practice, and his sister, an infectious disease nurse. His parents, who are enormously proud of their son, visit several times a year. His sister's preteen daughter, Mala, a beautiful, vibrant young girl, is like a daughter to Jessen. He spends much of his time with her. What Jessen does not have is a partner to share his life with. No one has yet compared to Andrew, the one who got away, the inspiration behind the Berlin patient. Jessen fills the void with close friends, his medical practice, and his family.

On a warm summer night in Berlin, I sit with Jessen on a rooftop terrace overlooking the city. He asks, "Do you think I should look at hydroxyurea again? Should I revisit the patients I gave it to?"

I nod my head. "You never know what you'll find."

The capital spreads out beneath us, a glittering masterpiece of contrasting architecture, on one side the stark modern buildings of the East, on the other the historic, ornate buildings of the West. The world of HIV once seemed as hopeless as the healing of Europe's twentieth-century wounds. But that night the cure seemed to sit peacefully in each of our laps. . . . Patients keep fighting their personal battles; researchers keep fighting their institutional battles; doctors keep trying to bring the two groups together for their mutual benefit. And for all of us.

Notes

The vast majority of information and quotes in this book were obtained from personal interviews with the parties involved. To simplify the references of this book, all major scientific and popular press articles from each Berlin patient are grouped together before proceeding to specific chapter references.

Christian Hahn—Berlin Patient #1

Scientific Reports

The first report that hydroxyurea could be used against HIV was shown in cell culture by Lisziewicz and Lori in Gallo's lab: "Hydroxyurea as an inhibitor of human immunodeficiency virus-type 1 replication," *Science* 266 (Nov 4, 1994).

An early report on the first Berlin patient by Lisziewicz, Lori, and Jessen: "HIV-1 suppression by early treatment with hydroxyurea, didanosine, and a protease inhibitor," *Lancet* 352 (Jul 18, 1998).

The principal report detailing the science of the first Berlin patient can be found in "Control of HIV despite the discontinuation of antiretroviral therapy," *New England Journal of Medicine* 340 (May 27, 1999).

Popular Press

The Mark Schoofs article interviewing the Berlin patient is "The Berlin Patient," *New York Times Magazine* (June 21, 1998).

The article that proved so contentious among Jessen and his colleagues is "Ray of Hope in the AIDS War," *Newsweek* (February 23, 1998).

The tabloid article that upset Jessen due to its bold use of the word *cure* is "AIDS die erste heilung?" *B.Z.* (June 18, 2000).

The story of the first Berlin patient was also covered in:

"HIV Suppressed Long after Treatment," *Science* (September 26, 1997).
"HIV Hope in Old Cancer Drug," *The Observer* (January 7, 1998).
"Der Berlin-Patient," *Rheinische Post* (October 9, 2004).
"Das medizinische wunder," *Tagesspiegel* (September 3, 2004).

Timothy Ray Brown—Berlin Patient #2

Scientific Reports
Gero Hütter first reported the Berlin patient in a poster session: "Treatment of HIV-1 infection by allogeneic CCR5-Δ32/Δ32 stem cell transplantation: a promising approach," Abstract 719, 15th Conference on Retroviruses and Opportunistic Infections, Boston, MA (2008).

Hütter first published the Berlin patient data in "Long-term control of HIV by CCR5 Delta32/Delta32 stem-cell transplantation," *New England Journal of Medicine* 360 (Feb 12, 2009).

Follow-up on the Berlin patient by Hütter was reported in "Eradication of HIV by transplantation of CCR5-deficient hematopoietic stem cells," *Scientific World Journal* 11 (May 5, 2011); "Evidence for the cure of HIV infection by CCR5 Δ32/Δ32 stem cell transplantation," *Blood* 117 (Mar 10, 2011); "The CCR5-delta32 polymorphism as a model to study host adaptation against infectious diseases and to develop new treatment strategies," *Experimental Biology and Medicine* 236 (Aug 2011); "Transplantation of selected or transgenic blood stem cells—a future treatment for HIV/AIDS?" *Journal of the International AIDS Society* 12 (Jun 28, 2009); "The effect of the CCR5-delta32 deletion on global gene expression considering immune response and inflammation," *Journal of Inflammation* 8 (Jan 2011); "Allogeneic transplantation of CCR5-deficient progenitor cells in a patient with HIV infection: an update after 3 years and the search for patient no. 2," *AIDS* 25 (Jan 14, 2011).

Popular Press
It would be impossible to cite all popular press articles mentioning Timothy Brown. Instead, I've listed the key articles that influenced the research community and the public.

The Mark Schoofs article that convinced *The New England Journal of Medicine* to publish Gero Hütter's paper is "A Doctor, a Mutation and a Potential Cure for AIDS," *Wall Street Journal* (November 7, 2008).

Timothy was given a sizable fee for his interview in *Stern*, a popular German magazine. "Der Mann, der HIV besiegte," *Stern* (December 8, 2010).

"The Man Who Had HIV and Now Does Not," *New York Magazine* (May 29, 2011).

"The Emerging Race to Cure HIV Infections," *Science* (May 13, 2011).

Chapter 1: The Good Doctor in Denial

The 1993 March on Washington for Lesbian, Gay, and Bi Equal Rights and Liberation was one of the largest civil rights demonstrations in American history. Video from the march is archived at the C-SPAN video library. http://www.c-spanvideo.org/program/40062-1.

Stages of viral disease including clinical descriptions of prodromal and incubation periods for HIV can be found in *Clinical Infectious Disease*, edited by David Schlossberg (Cambridge University Press, 2008).

Bob Siliciano's first paper identifying a latent reservoir for HIV was "Identification of a reservoir for HIV-1 in patients on highly active antiretroviral therapy," *Science* 278 (Nov 14, 1997).

"You are stuck with the virus unless you get every last latently infected cell" was quoted from an interview with Bob Siliciano, published in "Come out, come out," *International AIDS Vaccine Initiative Report* 9 (2005).

Eighty-three percent of physicians have prescribed medications for a family member was reported in "When physicians treat members of their own families," *New England Journal of Medicine* 325 (1991).

A beautiful description of the fall of the Berlin Wall and its global impact can be found in *1989: The Struggle to Create Post–Cold War Europe* by Mary Elise Sarotte (Princeton University Press, 2009).

Incredibly, there are still squatters from reunification in East Berlin. Their stories and many more can be found in a collection archived by National Public Radio: http://berlinstories.org.

Chapter 2: A Visit with the Family Doctor

The first report of a nucleic acid test to diagnose HIV in those with a negative antibody test, identical to the test used to diagnose Christian with HIV, was "Identification of HIV-infected seronegative individuals by a direct diagnostic test based on hybridisation to amplified viral DNA," *Lancet* 2 (1988).

David Ho was named *Time's* Man of the Year and featured on the cover of the magazine on December 30, 1996.

Roche discussed their new approach to targeting HIV's protease enzyme in "Rational design of peptide-based HIV proteinase inhibitors," *Science* 248 (1990).

Roche's saquinavir was approved for use by the FDA on December 6, 1995 while Merck's indinavir was approved roughly three months later on March 13, 1996.

Chapter 3: Death Sentence?

On page 2 of Christopher Isherwood's memoir, *Christopher and His Kind* (1976), the author says, speaking of himself in the third person, "To Christopher, Berlin meant Boys."

Further explanation of how HIV antibody tests work is available from "HIV assays: operational characteristics," *World Health Organization* (2002).

Detailed description of the innate and acquired immune system can be found in *Immunobiology*, 5th edition, *The Immune System in Health and Disease*, by Charles A Janeway Jr, Paul Travers, Mark Walport, and Mark J Shlomchik (Garland, 2001).

The CDC estimate that one-third of those diagnosed with HIV in the mid-1990s didn't return for their test results was reported in Centers for Disease Control and Prevention, "Advancing HIV prevention: new strategies for a changing epidemic—United States, 2003," *Morbidity and Mortality Weekly Report* 52 (2003).

More on OraQuick and other rapid HIV antibody tests is discussed in "A rapid review of rapid HIV antibody tests," *Current Infectious Disease Reports* 8 (2006).

Perspective on HIV as a death sentence in the 1990s is described in "HIV: Now and Then," *Gay Times* 415 (February 2013).

Chapter 4: Viral Trojan Horse

David Barry's interest in HIV is described in "The Inside Story of the AIDS Drug," *Fortune* (November 5, 1990), and in his obituary, "David Barry: Key Researcher in the Development of AZT," *Guardian* (March 8, 2002).

Gallo's report of the HTLV-III virus causing AIDS was in "A pathogenic retrovirus (HTLV-III) linked to AIDS," *New England Journal of Medicine* 311 (Nov 15, 1984).

A description of retroviruses and their life cycle can be found in chapter 3 of *Retroviruses: Molecular Biology, Genomics and Pathogenesis*, edited by Reinhard Kurth and Norbert Bannert (Caister Academic Press, 2010).

The extraterrestrial origin of uracil was described in "The Surface Composition of Titan," *American Astronomical Society, DPS Meeting* (March 2012).

Further discussion of viruses and their taxonomy can be found in *A Planet of Viruses* by Carl Zimmer (University of Chicago Press, 2011).

The long evolutionary history of FIV in cougars is described in "The molecular biology and evolution of feline immunodeficiency viruses of cougars," *Veterinary Immunology and Immunopathology* 123 (2008).

Estimates of the long evolutionary history that African green monkeys share with their SIV can be found in "SIVagm infection in wild African green monkeys from South Africa: epidemiology, natural history, and evolutionary considerations," *PLoS Pathogens* 9 (2012) and "Island biogeography reveals the deep history of SIV," *Science* 329 (2010).

HIV diversity and its relationship to drug resistance is further discussed in "HIV drug resistance," *New England Journal of Medicine* 350 (2004).

David Balitmore's Nobel Prize–winning work on reverse transcriptase can be found in "Reversal of information flow in the growth of RNA tumor viruses," *New England Journal of Medicine* 284 (1971).

How combination antiviral therapy (known as HAART) changed the world of HIV therapy in 1996 can be found in "The art of 'HAART': researchers probe the potential and limits of aggressive HIV treatments," *Journal of the American Medical Association* 277 (Feb 26, 1997).

Chapter 5: A Weapon from the War on Cancer

Speculation on whether World War II delayed early investigation into the environmental factors influencing cancer is discussed in "Historical threads in the development of oncology social work," *Journal of Psychosocial Oncology* 27 (2009).

Details on the early stigma of cancer and the biographical details of Mary Lasker can be found in *The Mary Lasker Papers* archived at the National Institutes of Health. Quotes from Mary Lasker were taken from twenty years of recordings of interviews between John T. Mason and Mary Lasker archived by Columbia University and available digitally in their notable New Yorkers section: http://www.columbia.edu/cu/lweb/digital/collections/nny/laskerm/index.html.

How Mary Lasker convinced NBC to lift the ban on cancer is described in "A Tribute to Mary Lasker," *Cancer News* 48 (1994).

Jerome Horwitz's research paper first describing AZT was "Nucleosides. IX. The formation of 2',2'-unsaturated pyrimidine nucleosides via a novel beta-elimination reaction," *Journal of Organic Chemistry* 31 (1966).

The definitive book on how DNA replicates is *DNA Replication*, 2nd edition, by Arthur Kornberg (University Science Books, 1992).

Background information on Dr. Jerome Horwitz was obtained from the following sources as well as interviews with colleagues and family members. Sadly, Horwitz passed away on September 6, 2012.

"The Inside Story of the AIDS Drug," *Fortune* (November 5, 1990).

"The Story of AZT: Partnership and Conflict," *Scribd* (2006).

"We had a very interesting set of compounds waiting for the right disease" is quoted from Horwitz in "A Failure Led to Drug Against AIDS," *New York Times* (September 20, 1986).

Quotes from Samuel Broder and background information obtained from the NIH online archive: "In their own words, NIH researchers recall the early years of AIDS," http://history.nih.gov/nihinownwords/index.html and a paper he authored: "The development of antiretroviral therapy and its impact on the HIV-1/AIDS pandemic," *Antiviral Research* 85 (2010).

Details of Gallo's press conference in 1984 were obtained from personal interviews and from the book *Virus Hunting: Aids, Cancer, and the Human Retrovirus: A Story of Scientific Discovery* by Robert Gallo (1993). Additional details were gained from an interview with Margaret Heckler included in the PBS *Frontline* program *"The Age of AIDS"* (2006).

Gallo and his team used HTLV-III as the name for what is now HIV in "Frequent detection and isolation of cytopathic references (HTLV-III) from patients with AIDS and at risk for AIDS," *Science* 224 (1984).

Anthony Fauci's quote "What lifestyle did the fetus undertake to acquire the disease?" is from the NIH online archive: "In their own words, NIH researchers recall the early years of AIDS," http://history.nih.gov/nihinownwords/index.html.

Many sources report the discrimination that those with HIV suffered during the 1980s and 1990s. These articles document a few specific means of discrimination:

"Ban on deadly kiss of life," *Sunday Mirror* (February 17, 1985).

"AIDS: prejudice and progress," *Time* (September 8, 1986).

"Voices: The miracle of Ryan White," *Time* (April 23, 1990).

Background on Janet Rideout and her role in AZT is discussed in "The Inside Story of the AIDS Drug," *Fortune* (November 5, 1990).

Description of protests including "Trials are treatment" obtained from interviews with former and current activists.

Chapter 6: The Days of Acting Up

Seventy percent of those born after 1980 favor same-sex marriage was reported by the Pew Research Center in the report "Growing Support for Gay Marriage: Changed Minds and Changing Demographics," (March 20, 2013).

Rent, written and composed by Jonathan Larson, became a Broadway production on April 29, 1996, at the Nederlander Theatre in NYC.

Report of the first promising signs using AZT for AIDS can be found in "AIDS therapy: first tentative signs of therapeutic promise," *Nature* 323 (1986).

The first paper describing AZT as an antiviral agent for a virus that would later become known as HIV is "3'-Azido-3'-deoxythymidine (BW A509U): an antiviral agent that inhibits the infectivity and cytopathic effect of human T-lymphotropic virus type III/lymphadenopathy-associated virus in vitro," *Proceedings of the National Academy of Sciences of the United States of America* 82 (1985).

Toxicity problems with AZT in AIDS patients was first reported in "The toxicity of azidothymidine (AZT) in the treatment of patients with AIDS and AIDS-related complex," *New England Journal of Medicine* 317 (1987). This paper lists the red blood cell transfusion rates for those taking AZT as 31 percent while those on a placebo had an 11 percent transfusion rate. This study also lists adverse reactions to AZT; 84 percent of those taking AZT reported an adverse reaction.

Results of the first AZT clinical trial were reported in "The efficacy of azidothymidine (AZT) in the treatment of patients with AIDS and AIDS-related complex, a double-blind, placebo-controlled trial," *New England Journal of Medicine* 317 (1987).

Effects of AZT in the bone marrow was discussed in Pluda JM, Mitsuya H, Yarchoan R, "Hematologic effects of AIDS therapies," *Hematology Oncology Clinics of North America* 5 (1991), and "Zidovudine pharmacokinetics in zidovudine-induced bone marrow toxicity," *British Journal of Clinical Pharmacology* 37 (1994).

Background on AZT development can be found in the author's Introduction to *North Carolina and the Problem of AIDS: Advocacy, Politics, and Race in the South* by Stephen Inrig (University of North Carolina Press, 2011).

"The debate over AZT clinical trials," Harvard University, John F. Kennedy School of Government, Case program (1999).

The Ethics and the Business of Bioscience by Margaret L. Eaton (Stanford Business Books, 2004).

Cost of AZT was reported in "AZT's Inhuman Cost," *New York Times* (August 28, 1989).

The $400 million in profits garnered by Burroughs Wellcome in 1992 was reported in "Market Place: Burroughs Wellcome, Analysts Say, Is More Than Just AZT," *New York Times* (June 10, 1993).

The cultural phenomena of AZT and AIDS activism is documented in The ACT UP Oral History Project, http://www.actuporalhistory.org/interviews/index.html.

Before the patent to AZT ended, in 2002 Jerome Horwitz joined a lawsuit against GlaxoSmithKline, formerly Burroughs Wellcome, for patent rights to AZT. Details about AZT development can be found in documentation from this lawsuit against GlaxoSmithKline in the United States Federal Court for Central District of California (Western Division, Case No. 02-5223 TJH Ex) by AIDS Healthcare Foundation.

Samuel Broder discussed how new drugs coming from the NCI were "an antidote to the sense of therapeutic nihilism" in an interview recorded in "In their own words, NIH researchers recall the early years of AIDS," http://history.nih.gov/nihinownwords/index.html.

"The perfect is the enemy of the good" is translated from the French in Voltaire's poem "La Bégueule."

Chapter 7: Recognizing a Global Pandemic

David Ho's history was obtained from personal interviews.

The first reported cases of an unknown disease that would later be identified as HIV were 5 homosexual men with biopsy-confirmed *Pneumocystis carinii* pneumonia from three hospitals in Los Angeles. This first report was "*Pneumocystis* pneumonia—Los Angeles," *Morbidity and Mortality Weekly Report* 30 (1981).

More details on smallpox can be found in *Smallpox: The Death of a Disease: The Inside Story of Eradicating a Worldwide Killer* by D. A. Henderson and Richard Preston (Prometheus, 2009).

David Ho's paper "Time to hit HIV, early and hard," *New England Journal of Medicine* 333 (1995), would have an enormous influence on the HIV community and inspire part of Heiko Jessen's therapy for Christian.

David Ho was named *Time's* Man of the Year and featured on the cover of the magazine on December 30, 1996.

Ho's work on combination therapy was featured in the cover story "The End of AIDS?" *Newsweek* (December 1, 1996).

GRID, or gay-related immune deficiency, was a term coined by the media. The term was both inaccurate and offensive since there is no link between homosexuality and the disease. In 1982, the CDC established the term acquired immune deficiency, or AIDS. This is detailed in "What to call the AIDS virus?" *Nature* 321 (1986).

More detail about the human leukocyte antigen and its role in the immune system can be found in *Immunobiology*, 5th edition, *The Immune System in Health and Disease*, by Charles A Janeway Jr, Paul Travers, Mark Walport, and Mark J Shlomchik (Garland, 2001).

Bruce Walker's personal history was obtained by personal interviews.

Walker's first paper was "HIV-specific cytotoxic T lymphocytes in seropositive individuals," *Nature* 328 (1987).

Chapter 8: From the One Percent

Notes from the 1993 International Conference on AIDS can be found in "We are all Berliners: notes from the Ninth International Conference on AIDS," *American Journal of Public Health* 83 (1993).

A nice review of CD4 T cells and their role in the immune system can be found in "CD4 T cells: fates, functions, and faults," *Blood* 112 (2008).

Perspective on the CONCORDE trial can be found in "After Concorde," *British Medical Journal* 306 (1993).

"How much is Roche paying you to say this?" is attributed to Margaret Fischl in "Once We Were Warriors: Activist Corpses Borne in Protest, Furtive Legislative Coups and the Devastation That Was Berlin," *Treatment Action Group* (2002). This article also describes the 1993 conference as the "most depressing AIDS conference ever."

Abstracts and data presented at the Ninth International AIDS Conference in Berlin in 1993 can be found on the AIDS Education Global Information System website: http://www.aegis.org/DisaplayConf/directory.aspx?Conf=The%20International%20AIDS%20Society-IAS.

Saquinavir was first described by Roche in "Antiviral properties of Ro 31-8959, an inhibitor of human immunodeficiency virus (HIV) proteinase," *Antiviral Research* 16 (1991).

Details of saquinavir development at Roche can be found in *Ethics and the Business of Bioscience* by Margaret L. Eaton (Stanford University Press, 2004).

Merck's early (and incorrect) structure of protease was reported in "Three-dimensional structure of aspartyl protease from human immunodeficiency virus HIV-1," *Nature* 337 (1989).

The effectiveness of HAART was reported in "Long term effectiveness of potent antiretroviral therapy in preventing AIDS and death: A prospective cohort study,"*Lancet* 366 (2005).

Two papers reported that HAART reduces death by 60–80 percent: "A controlled trial of two nucleoside analogues plus indinavir in persons with human immunodeficiency virus infection and CD4 cell counts of 200 per cubic millimeter or less," *New England Journal of Medicine* 337 (1997), and "Treatment with indinavir, zidovudine, and lamivudine in adults with human immunodeficiency virus infection and prior antiretroviral therapy," *New England Journal of Medicine* 337 (1997).

Chapter 9: But, Doctor, I Don't Feel Sick

Descriptions of HIV-positive patients at St. Clare's hospital were made from personal observations.

An overview of the controversy of when to begin antiviral therapy for HIV can be found in "When to start antiretroviral therapy—ready when you are?" *New England Journal of Medicine* 360 (2009).

The finding that HIV infects as a single "founder virus" was surprising in the field. Some believe characterization of this founder virus may lead to new vaccine strategies. This was first reported in "Identification and characterization of transmitted and early founder virus envelopes in primary HIV-1 infection," *Proceedings of the National Academy of Sciences* 105 (2008).

How T cells are infected by HIV and the mechanism of their destruction is described in "HIV preferentially infects HIV-specific CD4+ T cells," *Nature* 417 (2002).

Most HIV replication occurs in the intestines. This is discussed in "Getting to the guts of HIV pathogenesis," *Journal of Experimental Medicine* 200 (2004).

Numerous studies have examined the dense network of lymphocytes in the intestines and other mucosal tissue, including "Overview of the mucosal immune system," *Current Topics in Microbiology and Immunology* 146 (1989).

A discussion of the importance of mucosal lymphocytes in HIV infection can be found in "HIV pathogenesis: the first cut is the deepest," *Nature Immunology* 6 (2005).

Surprisingly, the kinetics of T cell depletion remains the same whether infection occurs through the mucosal route (rectal or vaginal) or intravenously.

This was described in "Gastrointestinal tract as a major site of CD4+ T cell depletion and viral replication in SIV infection," *Science* 280 (1998).

Numerous papers have described the destruction of T lymphocytes in the gut following HIV. One of the first is "Severe CD4+ T-cell depletion in gut lymphoid tissue during primary human immunodeficiency virus type 1 infection and substantial delay in restoration following highly active antiretroviral therapy," *Journal of Virology* 77 (2003).

The first report of the importance of CCR5 in HIV entry is "Identification of a major co-receptor for primary isolates of HIV-1," *Nature* 381 (1996).

How the HIV envelope protein fuses with the human cell is described in detail in "The HIV Env-mediated fusion reaction," *Biomembranes* 1614 (2003).

The number of CD4 T cells in the blood of an average person varies quite a bit, accounting for the range in values cited. This variation with average numbers controlled for sex and age is reported in "Laboratory control values for CD4 and CD8 T lymphocytes: implications for HIV-1 diagnosis," *Clinical Experimental Immunology* 88 (1992).

HIV makes on average 10 billion copies of itself a day as reported in "HIV-1 dynamics in vivo: Virion Clearance Rate, Infected Cell Life-Span, and Viral Generation Time," *Science* 271 (1996).

HIV is able to inflict massive cell death by direct cell killing and other indirect mechanisms. This cell death is discussed in detail in *Cell Death during HIV Infection*, edited by Andrew D. Badley (CRC Press, 2006).

Stopping antiviral therapy leads to mutation of the virus and development of drug resistance as reported in "Basic science kinetics of HIV-1 RNA and resistance-associated mutations after cessation of antiretroviral combination therapy," *AIDS* 15 (2001).

Chapter 10: The Delta 32 Mutation

The first paper detailing HIV resistance in individuals with a defect in their CCR5 gene was "Homozygous defect in HIV-1 co-receptor accounts for resistance of some multiply-exposed individuals to HIV-1 infection," *Cell* 86 (1996).

While most people with the Δ32 mutation lead healthy lives, a few studies have found that those without a working CCR5 gene are at higher risk for West Nile virus. These studies are controversial. We can't be certain of all the potential consequences of CCR5 deficiency. Of interest, other studies have pointed to protection conferred by the Δ32 in other disease, such as cerebral malaria, from the mutation. Both sides can be found in "CCR5 deficiency increases risk of symptomatic West Nile virus infection," *Journal of Experimental Medicine* 203 (2006); "CCR5 deficiency is a risk factor for early clinical manifestations of West Nile virus infection but not for viral transmission," *Journal of Infectious Diseases* 201 (2010); "Role of chemokines polymorphisms in diseases," *Immunology Letters* 145 (2012).

The vast majority of transmitted viruses use the CCR5 receptor to enter cells, as reported in David Ho's paper "Genotypic and phenotypic characterization of HIV-1 patients with primary infection," *Science* 261 (1993).

Viruses use CCR5 no matter how they were originally transmitted, by sex, intravenously, or passed from mother to child: "Macrophage-tropic variants initiate human immunodeficiency virus type 1 infection after sexual, parenteral, and vertical transmission," *Journal of Clinical Investigation* 94 (1994).

The CCR5 Δ32 mutation is commonly found in Western European populations as reported in "Resistance to HIV-1 infection in Caucasian individuals bearing mutant alleles of the CCR-5 chemokine receptor gene," *Nature* 382 (1996); "The geographic spread of the CCR5 Delta32 HIV-resistance allele," *PLoS Biology* 3 (2005).

The paper that Hütter read in his medical school library and the first paper to describe the Δ32 mutation and its link to HIV was "Resistance to HIV-1 infection in Caucasian individuals bearing mutant alleles of the CCR-5 chemokine receptor gene," *Nature* 382 (1996).

Chapter 11: Calling All Elite Controllers

A Song in the Night: A Memoir of Resilience by Bob Massie (Doubleday, 2012).

Details on Bruce Walker's relationship with Bob Massie and the development of a cohort of elite controllers were obtained by personal interviews. Additional details and the quote "I must have audibly gasped" obtained from "Secrets of the HIV controllers," *Scientific American* 307 (2012).

Gene therapy for cystic fibrosis was described in "Cystic fibrosis transmembrane conductance regulator protein repair as a therapeutic strategy in cystic fibrosis," *Current Opinion in Pulmonary Medicine* 16 (2010).

Gene therapy for Parkinson's disease was described in "Safety and tolerability of gene therapy with an adeno-associated virus (AAV) borne GAD gene for Parkinson's disease: an open label, phase I trial," *Lancet* 369 (2007).

Gene therapy for Beta-thalassemia was described in "Beta-thalassemia treatment succeeds, with a caveat," *Science* 326 (2009).

Gene therapy for inherited blindness was described in Maguire AM, High KA, Auricchio A, Wright JF, Pierce EA, Testa F, Mingozzi F, Bennicelli JL, Ying GS, Rossi S, et al., "Age-dependent effects of RPE65 gene therapy for Leber's congenital amaurosis: a phase 1 dose-escalation trial," *Lancet* 374 (9701):1597-1605; 2009.

Details on Jesse Gelsinger's death after receiving an experimental gene therapy in 1999 and its impact on research is discussed in "Gene therapy death prompts review of adenovirus vector," *Science* 286 (1999).

Epidemiology and description of HIV controllers can be found in "Prevalence and comparative characteristics of long-term nonprogressors and HIV controller patients in the French Hospital Database on HIV," *AIDS* 23 (2009).

The high expression of CCR5 in tissue lymphocytes is described in "Expression of the chemokine receptors CCR4, CCR5, and CXCR3 by human tissue-infiltrating lymphocytes," *American Journal of Pathology* 160 (2002).

The importance of the gut in acute HIV is discussed in "Immunopathogenesis of acute AIDS virus infection," *Current Opinion in Immunology* 18 (2006).

An overview of HLA and HIV written by Walker is described in "HIV and HLA class I: an evolving relationship," *Immunity* 37 (2012).

The HLA alleles B*27 and B*57 are found frequently in HIV controllers as described in "HLA alleles associated with delayed progression to AIDS contribute strongly to the initial CD8+ T-cell response against HIV-1," *PLoS Medicine* 3 (2006).

The monkey version of elite HIV control, the Mamu-A*01 allele, was shown to protect primates from SIV in "Mamu-A*01 allele-mediated attenuation of disease progression in simian-human immunodeficiency virus infection," *Journal of Virology* 76 (2002).

Specific amino acids in the groove of the HLA-B gene that confer HIV control was described in "The major genetic determinants of HIV-1 control affect HLA class I peptide presentation," *Science* 330 (2010).

The association of HLA-B*57 and psoriasis was reported in "HLA-B57 is significantly associated with psoriasis in Northeast Romania," *Roumanian Archives of Microbiology and Immunology* 61 (2002).

Chapter 12: Treatment in Hiding

Percent of medical school graduates entering family medicine residencies was reported in "Entry of US medical school graduates into family medicine residencies," *Family Medicine* 44 (2012).

Development of hydroxyurea for clinical use was first published in "Hydroxyurea. A new type of potential antitumor agent," *Journal of Medicinal Chemistry* 6 (1963).

Basic mechanism of action, and diseases hydroxyurea is FDA-approved for, can be found in the product insert for Hydrea, hydroxyurea capsules, USP, from Bristol-Myers Squibb.

Chapter 13: The Second Diagnosis

The 25 percent 5-year survival rate for adult patients in the US from 2006 to 2010 diagnosed with AML and 11 percent 5-year survival rate for those patients experiencing a relapse is reported by the National Cancer Institute in the *SEER cancer statistics review, 1975–2010* (2012).

A review of AML, how it invades tissue and its clinical ramifications can be found in "Acute myeloid leukaemia in adults," *Lancet* 381 (2013).

Discussion of the role immune suppression might play in AML can be found in "Commentary: does immune suppression increase risk of developing acute myeloid leukemia?" *Leukemia* 26 (2012).

More detail on hematopoietic stem cells can be found in *Hematopoietic Stem Cell Biology*, edited by Motonari Kondo (Humana Press, 2010).

Allogeneic stem cell transplantation is described in "Allogeneic hematopoietic cell transplantation for acute myeloid leukemia when a matched related donor is not available," *Hematology* (2008).

CXCR4 isolates of HIV are expressed later in infection as described in "The HIV co-receptors CXCR4 and CCR5 are differentially expressed and regulated on human T lymphocytes," *Proceedings of the National Academy of Sciences of the United States of America* 94 (1997).

The pathogenic nature of the CXCR4 virus was described in "Phenotypic and genotypic comparisons of CCR5- and CXCR4-tropic human immunodeficiency virus type 1 biological clones isolated from subtype C-infected individuals," *Journal of Virology* 78 (2004).

Murine fetuses that lack CXCR4 are lethal, as reported in "Mechanism of human stem cell migration and repopulation of NOD/SCID and B2mnull NOD/SCID mice. The role of SDF-1/CXCR4 interactions," *Annals of the New York Academy of Sciences* 938 (2001), whereas mice that lack CCR5 survive, as reported in "Mice with a selective deletion of the CC chemokine receptors 5 or 2 are protected from dextran sodium sulfate-mediated colitis: lack of CC chemokine receptor 5 expression results in a NK1.1+ lymphocyte-associated Th2-type immune response in the intestine," *Journal of Immunology* 164 (2000).

A review of graft-versus-host disease can be found in "Concise review: acute graft-versus-host disease: immunobiology, prevention, and treatment," *Stem Cells Translational Medicine* 2 (2013).

More information about the ZKRD and rates of stem cell transplants in Germany can be found at the website http://www.zkrd.de/en/index.php

Likelihood of bone marrow match success in the United States calculated from data and analysis by the national marrow donor program and can be obtained from the website http://marrow.org/Home.aspx.

Age of the Δ32 mutation and postulation that its high frequency in Western European populations is linked to the bubonic plague was first proposed in the paper "Dating the origin of the CCR5-Delta32 AIDS-resistance allele by the coalescence of haplotypes," *American Journal of Human Genetics* 62 (1998). This theory is controversial, and other studies attempting to replicate this selective pressure in CCR5-deficient mice have been conflicting. See "Evolutionary genetics: CCR5 mutation and plague protection," *Nature* 427 (2004); "Evolutionary genetics: ambiguous role of CCR5 in Y. pestis infection," *Nature* 430 (2004); "The evolutionary history of the CCR5-Delta32 HIV-resistance mutation," *Microbes and Infection* 7 (2005); "The Black Death and AIDS: CCR5 Δ32 in genetics and history," *Quarterly Journal of Medicine* 99 (2006).

Description of one of the oldest mutations in humans can be found in "Adaptive Evolution of the FADS Gene Cluster within Africa," *PLoS One* 9 (2012).

Chapter 14: The Compassionate Use Exception

The compassionate use exception is described in congressional testimony from the U.S. Food and Drug Administration's statement *Availability of Investigational Drugs for Compassionate Use* by Robert Temple (June 20, 2001).

The statistic, 44 percent of HIV-positive gay men in the United States don't know they are infected, comes from a CDC survey: "Prevalence and awareness of HIV infection among men who have sex with men—21 cities, United States, 2008," *Morbidity and Mortality Weekly Report* 59 (2010).

Chapter 15: Three Deadly Diseases Move In

A hilarious description of the difficulty of taking the antiviral drug DDI (didanosine) can be found in *Queer and Loathing: Rants and Raves of a Raging AIDS Clone* by David B. Feinberg (Penguin Books, 1995).

Bioavailability data for didanosine, brand-name Videx, can be found in Table 10 of the package insert from Bristol-Myers Squibb.

Chapter 16: The Comfort of Family and Strangers

Studies examining the benefits of supportive family on the health of lesbian, gay, and bisexual individuals include: "The health of people classified as lesbian, gay and bisexual attending family practitioners in London: a controlled study," *BMC Public Health* 6 (2006); "Family rejection as a predictor of negative health outcomes in white and Latino lesbian, gay, and bisexual young adults," *Pediatrics* 123 (2009); "Parents' supportive reactions to sexual orientation disclosure associated with better health: results from a population-based survey of LGB adults in Massachusetts," *Journal of Homosexuality* 59 (2012); "A qualitative exploration of sexual risk and HIV testing behaviors among men who have sex with men in Beirut, Lebanon," *PLoS ONE* 7 (2012).

HIV and cachexia are further discussed in "HIV-related cachexia: potential mechanisms and treatment," *Oncology* 49 (1992).

The links between mitochondria, antiretroviral therapy and lipoatrophy are reported in "Mitochondrial RNA and DNA alterations in HIV lipoatrophy are linked to antiretroviral therapy and not to HIV infection," *Antiviral Therapy* 13 (2008).

Chapter 17: Timing

All quotes and background information from Dr. Julianna Lisziewicz obtained by personal interview.

A description of how mRNA is translated can be found in *The Cell: A Molecular Approach*, 2nd edition, by Geoffrey Cooper (Sinauer Associates, 2000).

The gene therapy approaches pioneered by Julianna Lisziewicz can be found in her paper "Gene therapy approaches to HIV infection," *American Journal of Pharmacogenomics* 2 (2002).

More information on the research institute founded by Lisziewicz and Lori can be found at the RIGHT website: http://www.rightinstitute.net.

Chapter 18: Transplanting

A review of who should get stem cell transplants for AML, and why, can be found in "Who should be transplanted for AML?" *Leukemia* 15 (2001).

Data for the low survival rate after a second transplant for AML can be found in "Prognosis of patients with a second relapse of acute myeloid leukemia," *Leukemia* 14 (2000). The 11 percent 5-year survival rate in adults with a relapse of AML is reported in "Prognostic index for adult patients with acute myeloid leukemia in first relapse," *American Society of Clinical Oncology* 23 (2005).

How conditioning regimens dampen the immune system in AML transplants can be found in "Myeloablative conditioning regimens for AML allografts: 30 years later," *Bone Marrow Transplantation* 32 (2003).

Chapter 19: "Perhaps We Have Eradicated HIV"

Details on how CD4:CD8 ratios are used by clinicians can be found in "CD4 percentage, CD4 number, and CD4:CD8 ratio in HIV infection: which to choose and how to use," *Journal of Acquired Immune Deficiency Syndromes* 2 (1989).

Siliciano's highly influential paper is "Identification of a reservoir for HIV-1 in patients on highly active antiretroviral therapy," *Science* 278 (1997).

Percentage of resting T cells in the blood and relationship to HIV is described in "Cellular APOBEC3G restricts HIV-1 infection in resting CD4+ T cells," *Nature* 435 (2005).

Lymph nodes are a favorite target for HIV. This is described in "Lymph node pathology of acquired immunodeficiency syndrome (AIDS)," *Annals of Clinical and Laboratory Science* 20 (1990).

Cecil Fox's paper that influenced his involvement in both Berlin patients is "HIV in infected lymph nodes," *Nature* 370 (1994).

Destruction of lymph node architecture is discussed in "Human immunodeficiency virus pathogenesis: insights from studies of lymphoid cells and tissues," *Clinical Infectious Disease* 33 (2001).

Walker's influential papers discussed from 1996 include "Recognition of the highly conserved YMDD region in the human immunodeficiency virus type 1 reverse transcriptase by HLA-A2-restricted cytotoxic T lymphocytes from an asymptomatic long-term nonprogressor," *Journal of Infectious Diseases* 173 (1996); "T cell receptor usage and fine specificity of human immunodeficiency virus 1-specific cytotoxic T lymphocyte clones: analysis of quasispecies recognition reveals a dominant response directed against a minor in vivo variant," *Journal of Experimental Medicine* 183 (1996); "Strong cytotoxic T cell and weak neutralizing antibody responses in a subset of persons with stable nonprogressing HIV type 1 infection," *AIDS Research and Human Retroviruses* 12 (1996); "Cytotoxic T lymphocytes in asymptomatic long-term nonprogressing HIV-1 infection. Breadth and specificity of the response and relation to in vivo viral quasispecies in a person with prolonged infection and low viral load," *Journal of Immunology* 156 (1996); "Efficient lysis of human immunodeficiency virus type 1-infected cells by cytotoxic T lymphocytes," *Journal of Virology* 70 (1996).

The interaction of HIV and IFN-γ in the ELISPOT assay is described in "The role of IFN-[gamma] Elispot assay in HIV vaccine research," *Nature* 4 (2009).

The debate over problems with the use of ELISPOT in HIV is described in "The role of IFN-g Elispot assay in HIV research," *Nature Protocols* 4 (2009).

Chapter 20: An Unexciting Recovery

Percent GDP which fuels welfare in Germany compared to the United States was obtained from "What the European and American welfare states have in common and where they differ: facts and fiction in comparisons of the European Social Model and the United States," *Journal of European Social Policy* 20 (2010).

The CROI talk in 2008 that released results from MOTIVATE on the drug maraviroc was "Efficacy and safety of maraviroc plus optimized background therapy in treatment-experienced patients infected with CCR5-tropic HIV-1: 48-week combined analysis of the MOTIVATE studies," Abstract #792, 15th Conference on Retroviruses and Opportunistic Infections, Boston, MA (2008).

Chapter 21: Trials

Walker's early paper looking at immune responses in acute HIV is "Vigorous HIV-1-specific CD4+ T cell responses associated with control of viremia," *Science* 278 (1997).

The ACTG 5025 trial was also known as "A study of the safety and effectiveness of hydroxyurea in patients on potent antiretroviral therapy and who have less than 200 copies/ml of HIV RNA in their blood." Details on the investigators, drugs given, and number of patients recruited can be found, like all clinical trials in the United States, through clinicaltrials.gov.

The saga of ACTG 5025 is covered in "Pancreatitis Deaths Shut Down ACTG 5025," *HIV Plus Magazine* (February/March 2000).

Differing perspective on hydroxyurea and clinical trials from Julianna Lisziewicz and Franco Lori can be found in this review written by them: "Hydroxyurea in the treatment of HIV infection: clinical efficacy and safety concerns," *Drug Safety* 26 (2003).

The warning letter sent by the FDA to Bristol-Myers Squibb on the marketing of hydroxyurea and false claims of ACTG 5025 was sent on October 27, 1999, and can be found on the FDA website at http://www.fda.gov/downloads/Drugs/GuidanceComplianceRegulatoryInformation/EnforcementActivitiesbyFDA/WarningLettersandNoticeofViolationLetterstoPharmaceuticalCompanies/UCM166219.pdf.

Jerome Horwitz's paper on discovering d4T is "Nucleosides. X. The action of sodium ethoxide on 3'-0-tosyl-2'-deoxyadenosine," *Tetrahedron Letters* 7 (13) (1966).

History of d4T and Yale University can be found in "Yale Pressed to Help Cut Drug Costs in Africa," *New York Times* (March 12, 2001).

The incidence of neuropathy in patients on d4T is discussed in "Human immunodeficiency virus-neuropathy with special reference to distal sensory polyneuropathy and toxic neuropathies," *Annals of Tropical Medicine and Public Health* 1 (2008).

Deborah Cotton was quoted as saying, "I'm not sure how good our advice was today," in "F.D.A. Panel Recommends AIDS Drug Despite Incomplete Data," *New York Times* (May 21, 1994).

Walker's results using treatment interruptions at first seemed promising as reported in "Immune control of HIV-1 after early treatment of acute infection," *Nature* 407 (2000).

Fauci was quoted as saying, "The strategy needs to be tested. The stop-and-go game can lead to drug resistance even if it looks so far like the wild type strain remains" in "Absence Makes the HAART Grow Fonder," *The Body* (February 1999).

The SMART study that changed the prevailing view on treatment interruptions was "CD4+ count–guided interruption of antiretroviral treatment: the strategies for management of antiretroviral therapy (SMART) study group," *New England Journal of Medicine* 355 (2006).

Results from Lori and Lisziewicz's study of hydroxyurea were published in "Lowering the dose of hydroxyurea minimizes toxicity and maximizes anti-HIV potency," *AIDS Research and Human Retroviruses* 21 (2005).

Chapter 22: Proof of Principle

Jeffrey Laurence was quoted as saying, "I thought it was the most exciting thing I'd heard about since the discovery of the virus. I couldn't believe people didn't take notice," in "The Man Who Had HIV and Now Does Not," *New York Magazine* (May 29, 2011).

The Boston Think Tank that Gero Hütter attended in 2008 was written up by researchers Rowena Johnston and Jeffrey Laurence in "amFAR Think Tanks: A Blueprint for Action Against HIV/AIDS," *amFAR, The Foundation for AIDS Research Newsletter* (September 16, 2008).

John Zaia's three-pronged attack to HIV is described in "Safety and efficacy of a lentiviral vector containing three anti-HIV genes—CCR5 ribozyme, tat-rev siRNA, and TAR decoy—in SCID-hu mouse-derived T cells," *Molecular Therapy* 15 (2007).

Gene therapy in AIDS lymphoma patients at City of Hope by Zaia is described in "RNA-based therapy for HIV with lentiviral vector-modified CD34(+) cells in patients undergoing transplantation for AIDS-related lymphoma," *Science Translational Medicine* 2 (2010).

Chapter 23: The Good Doctor in Court

Up until 1991, methadone could be administered in Germany only to drug users with highly specific criteria (which included AIDS). Numerous family doctors ignored these regulations, prescribing the drug to opiate addicts. Germany prosecuted the doctors, taking away many licenses. The narcotics act in 1992, called Betäubungsmittelgesetz, legalized methadone. However, a special license is still required to distribute it. A longer discussion of the history of methadone in Germany can be found in "Substitution treatment for opiod addicts in Germany," *Harm Reduction Journal* 4 (2007).

Chapter 24: Not Even Surprising

The abstract book and meeting report for the Fourth International Workshop on HIV Persistence during Therapy, which includes details on Robert Gallo's opening talk and Gero Hütter's presentation, are available from the global antiviral journal at: http://www.ihlpress.com/gaj_persistence2009 .html.

Information from Anthony Fauci concerning his opinions on the Berlin patients was obtained from a personal interview.

Fauci was quoted as saying, "It's very nice, and it's not even surprising, but it's just off the table of practicality," in "Rare Treatment Is Reported to Cure AIDS Patient," *New York Times* (November 13, 2008).

The average lifetime cost of antiviral drugs is reported as $709,731 for undiscounted therapy and $425,440 with discounts in "Newer drugs and earlier treatment: impact on lifetime cost of care for HIV-infected adults," *AIDS* 26 (2012).

The 2011 Milliman Medical Index estimated the cost of an allogeneic bone marrow transplant like Timothy received to be $805,400 USD: http://publications.milliman.com/research/health-rr/pdfs/2011-us-organ-tissue.pdf.

Cost of Atripla is discussed in "Generic HIV drugs will widen US treatment net," *Nature* (August 15, 2012).

Neurophysiology of HIV and aging is reported in "Pathways to neurodegeneration: effects of HIV and aging on resting-state functional connectivity," *Neurology* (2013); "Where does HIV hide? A focus on the central nervous system," *Current opinion in HIV and AIDS* (2013).

The pervasiveness of neurological conditions in HIV is reported in "HIV-associated neurocognitive disorder: pathogenesis and therapeutic opportunities," *Journal of Neuroimmune Pharmacology* 5 (2010).

Life expectancy for HIV-positive individuals in developed countries has leaped in recent years. These gains have been reported in "Life expectancy of individuals on combination antiretroviral therapy in high-income countries: a collaborative analysis of 14 cohort studies," *Lancet* 372 (2008); "Potential gains in life expectancy from reducing heart disease, cancer, Alzheimer's disease, kidney disease or HIV/AIDS as major causes of death in the USA," *Public Health* (2013).

Rise of life expectancy for those beginning early antiretroviral therapy was reported in "Projected life expectancy of people with HIV according to timing of diagnosis," *AIDS* 26 (2012).

Chapter 25: The Promise Kept

Fire with Fire is a short film directed by Ross Kauffman © Red Light Films (2013). It can be found online at http://focusforwardfilms.com/films/72/fire-with-fire.

More information on the CART-19 trial at Children's Hospital of Philadelphia can be found at their websites: http://www.chop.edu/service/oncology/pediatric-cancer-research/t-cell-therapy.html; http://www.chop.edu/system/galleries/download/pdfs/articles/oncology/summit-grupp-cart19.pdf.

More information on Emma Whitehead's story can be found on her website, where her mother, Kari Whitehead, has chronicled their experiences: http://emilywhitehead.com, and in "In Girl's Last Hope, Altered Immune Cells Beat Leukemia," *New York Times* (December 9, 2012).

All background information on Carl June obtained from a personal interview.

Information on bone marrow transplants and the Cold War obtained from "Atomic Medicine: the Cold War Origins of Biological Research," *History Today* 59 (2009).

All background information on Bruce Levine obtained from a personal interview.

Written by June and Levine, this article discusses their approaches to CCR5 gene therapy: "Blocking HIV's attack," *Scientific American* 306 (2012).

The paper in which Levine and June looked at dendritic cells was published in 1996: "Antiviral effect and ex vivo CD4+ T cell proliferation in HIV-positive patients as a result of CD28 co-stimulation," *Science* 272 (1996).

Background information on Edward Lanphier was obtained from a personal interview.

Percent homology between HIV and SIV is reported in "Comparison of SIV and HIV-1 genomic RNA structures reveals impacct of sequence evolution on conserved and non-conserved structural motifs," *PLoS Pathogens* 9 (2012).

Further reading on the development and application of ZFNs can be found in "Zinc finger nucleases: custom-designed molecular scissors for genome engineering of plant and mammalian cells," *Nucleic Acids Research* 33 (2005).

Further reading on SIV and HIV can be found in "Where the wild things are: pathogenesis of SIV infection in African nonhuman primate hosts," *HIV/AIDS Reports* 7 (2010), and "Natural SIV hosts: showing AIDS the door," *Science* 335 (2012).

Evidence that SIV has evolved over 32,000 years was published in "Island biogeography reveals the deep history of SIV," *Science* 329 (2010).

Comparison of pathogenic and nonpathogenic monkey models of SIV can be found in "AIDS pathogenesis: a tale of two monkeys," *Journal of Medical Primatology* 37 (2008).

Genetic similarities between "Black 6" mice and humans is reported in "Of Mice and Men: Striking Similarities at the DNA Level Could Aid Research," *San Francisco Chronicle* (December 5, 2002).

Comparison of mouse and human gene expression was published in "Genomic responses in mouse models poorly mimic human inflammatory diseases," *Proceedings of the National Academy of Sciences of the United States of America* 110 (2013).

A review of humanized mice in HIV infection can be found in "Humanized mouse models of HIV infection," *AIDS Reviews* 13 (3):135-148 (2011).

Some researchers don't believe that humanized mice represent a viable model for HIV therapies. This perspective is presented in "The mouse is out of the bag: insights and perspectives on HIV-1-infected humanized mouse models," *Experimental Biology and Medicine* 236 (2011).

Carl June's test of CCR5 ZFNs in humanized mice challenged with HIV was published in "Establishment of HIV-1 resistance in CD4+ T cells by genome editing using zinc-finger nucleases," *Nature Biotechnology* 26 (2008).

Results from Carl June's CCR5 ZFN clinical trial in HIV-positive volunteers was presented at "HAART treatment interruption following adoptive transfer of zinc finger nuclease (ZFN) modified autologous CD4+ T-cells (SB-728-T) to HIV-infected subjects demonstrates durable engraftment and suppression of viral load," Abstract #165, 18th Conference on Retroviruses and Opportunistic Infections, Boston, MA (2011); "Induction of acquired CCR5 deficiency with zinc finger nuclease-modified autologous CD4 T cells (SB-728-T) correlates with increases in CD4 count and effects on viral load in HIV-infected subjects," Abstract #155, 19th Conference on Retroviruses and Opportunistic Infections, Seattle, WA (2012).

The 2011 Milliman Medical Index estimates the cost of an autologous transplant as performed by Carl June's team using CCR5 ZFNs to be $363,800. Therefore, the cost of an autologous transplant is approximately $300,000 less than that of lifetime antiviral therapy. http://publications. milliman.com/research/health-rr/pdfs/2011-us-organ-tissue.pdf.

Chapter 26: A Child Cured—So What?

Report of an HIV cure in a toddler was made in "Functional HIV cure after very early ART of an infected infant," Abstract #48LB, 20th Conference on Retroviruses and Opportunistic Infections, Atlanta, GA (2013).

Details on Eric and John Wagner's quote reported in "Revolutionary treatment begins," *University of Minnestota News* (April 24, 2013).

Timothy Brown's phone call to Eric Blue was reported in "Babies could be key to HIV cure," *Washington Blade* (Aplril 26, 2013).

Background information on David Margolis was obtained by personal interview.

History of HDACi in cancer as well as the approval of vorinostat as the first HDACi by the FDA are discussed in "Histone deacetylase (HDAC) inhibitors in recent clinical trials for cancer therapy," *Clinical Epigenetics* 1 (Dec 2010).

Investigation of the HDACi, valproic acid, by David Margolis can be found in "Coaxing HIV-1 from resting CD4 T cells: histone deacetylase inhibition allows latent viral expression," *AIDS* 18 (May 21, 2004); "Depletion of latent HIV-1 infection in vivo: a proof-of-concept study," *Lancet* 366 (Aug 13, 2005).

David Margolis's data on vorinostat can be found in "Expression of latent HIV induced by the potent HDAC inhibitor suberoylanilide hydroxamic acid," *AIDS Research and Human Retroviruses* 25 (Feb 2009).

A nice review on HDACi from David Margolis can be found in "Histone deacetylase inhibitors and HIV latency," *Current Opinion in HIV and AIDS* 6 (2011).

David Margolis's presentation to an overflowing crowd: "Administration of vorinostat disrupts HIV-1 latency in patients on ART," Abstract #157LB,

19th Conference on Retroviruses and Opportunistic Infections, Seattle, WA (2012).

Sharon Lewin shared her data and perspective on vorinostat in "HIV latency and eradication: clinical perspectives," Abstract #106, 19th Conference on Retroviruses and Opportunistic Infections, Seattle, WA (2012).

David Margolis was quoted as saying, "This proves for the first time that there are ways to specifically treat viral latency, the first step towards curing HIV infection," in a press release, "Drug helps purge hidden HIV virus, UNC study shows" from the University of North Carolina School of Medicine (March 8, 2012).

Chapter 27: Zinc Finger Snap

The paper from my dissertation research that Paula Cannon and I wrote is "Zinc finger nuclease-mediated CCR5 knockout hematopoietic stem cell transplantation controls HIV-1 in vivo," *Nature Biotechnology* 28 (Aug 2010).

Background information on Paula Cannon obtained from personal interviews.

CIRM funding for the "dream team" is reported online by CIRM at http://www.cirm.ca.gov/our-funding/awards/ziinc-finger-nuclease-based-stem-cell-therapy-aids.

Paula Cannon's quote "The fact that it worked, it was like the 'Duh!' moment. It was the most unremarkable thing. I wasn't quite prepared for how exciting people thought the results were" is from "Locking Out HIV," *CIRM Annual Report* (2011).

Background information on Timothy Henrich and Dan Kuritzkes obtained from personal interviews.

Tim Henrich first presented results from the Boston patients in "Long-term reduction in peripheral blood HIV-1 reservoirs following reduced-intensity conditioning allogeneic stem cell transplantation in two HIV-positive individuals," Abstract THAA0101, Ninteenth International AIDS Conference, Washington, DC (2012).

The Boston patients received wide coverage in the press, including "Two More Nearing AIDS 'Cure' after Bone Marrow Transplants, Doctors Say," *National Public Radio, Shots health blog* (July 26, 2012).

Results from the VISCONTI cohort are published in "Post-treatment HIV-1 controllers with a long-term virological remission after the interruption of early initiated antiretroviral therapy ANRS VISCONTI study," *PLoS Pathogens* (Mar 24, 2013).

Background information on David Baltimore was obtained from a personal interview.

David Baltimore and Irvin Chen's first paper on siRNA is "Inhibiting HIV-1 infection in human T cells by lentiviral-mediated delivery of small interfering RNA against CCR5," *Proceedings of the National Academy of Sciences of the USA* 100 (Jan 7, 2003).

"amFAR's interest in exploring the role of gene therapy in the eradication of HIV infection stems from a February 2009 report in the *New England Journal of Medicine* of a patient in Berlin" is from "Manipulating the Smallest Building Blocks of Life to Defeat the World's Biggest Infectious Disease Killer," *amFAR, The Foundation for AIDS Research press release* (February 18, 2010).

Irvin Chen and David Baltimore's $20 million grant from CIRM, "HPSC based therapy for HIV disease using RNAi to CCR5," was reported in "Researchers knock down gene to stop HIV in its tracks," *Nature Medicine* 16 (2010).

Chapter 28: The Abused, the Respected, the Relentless

Steve Yukl's talk on Timothy's remnant virus was "Increased risk of virologic rebound in patients on antiviral therapy with isolated detectable viral loads <48 copies/ml by Taqman PCR RT-PCR Assay," International Workshop on HIV & Hepatitis Virus Drug Resistance and Curative Strategies, Sitges, Spain (2012).

"If you do enough cycles of PCR, you can get a signal in water for pink elephants" was quoted in "Evidence That Man Cured of HIV Harbors Viral Remnants Triggers Confusion," *Science Insider* (June 11, 2012).

The press release by Alain Lafeuillade is "The So Called HIV Cured 'Berlin' Patient Still Has Detectable HIV in His Body," *PRWeb UK* (June 11, 2012).

"Despite the possibility of intermittently detectable, very low levels of HIV, the Berlin patient has remained off of ART for 5 years, has no detectable viremia using standard assays, has waning HIV antibody levels, has limited to undetectable HIV-specific T cell responses, and has no evidence of HIV-related immunologic progression. The patient certainly meets any clinical definition for having achieved a long-term remission, and may even have had a sterilizing cure. Even the most extraordinary 'elite' controllers described in the literature have more robust evidence for persistent infection." This is quoted from "Challenges in detecting HIV persistence during potentially curative interventions: a study of the Berlin patient," *PLoS Pathogens* (May 9, 2013). This paper also contains the virus data analyzed by two labs from Timothy's samples taken five years after he stopped therapy.

Timeline

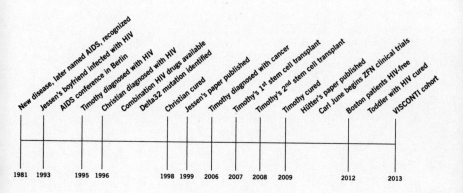

Acknowledgments

First and foremost, I would like to thank Timothy Ray Brown and Christian Hahn. Both men have given themselves to science and medicine in ways that most of us cannot contemplate. They have been generous with their stories, sharing details and experiences that are not easy to relive. Likewise, the friends, family, and partners of both Berlin patients have been incredibly generous in sharing their experiences and giving me a glimpse of what life is like when you share it with someone cured of HIV. Timothy's foundation can be found at worldaidsinstitute.org.

Similarly, this book would not exist without Heiko Jessen and Gero Hütter. Both of these men have influenced the lives of millions. I feel privileged to tell their personal stories and the science they've brought to the world. Their colleagues in Berlin, both at the Jessen clinic and at Charité hospital, have been incredibly kind and helpful.

This would never have become a book without my agent Laurie Abkemeier. Every step of the way she's been there, from fighting through the first murky drafts, to panicked e-mails from Berlin, to finding just the right publisher.

I'm forever indebted to my editor, Stephen Morrow. From our first conversation, Stephen has made this book better. He's taken dense immunology text and turned it into clear science prose. There isn't a page he hasn't improved. Stephanie Hitchcock, assistant editor, has brought much to the book through her thoughtful revisions and questions. Her passion for books shines through her edits and I feel fortunate to have her as a reader. The entire Dutton team has brought incomparable expertise to the project and I am very lucky to work with them.

I've been fortunate to receive tremendous help in the research of this book. I would be lost without Paula Cannon who has given so much of herself both professionally and personally. My education in humanized mice came from Victor Garcia and Paul Denton. I thank Gay Crooks, Don Kohn, and the members of their respective laboratories for sharing their expertise in stem cells and gene therapy.

Bruce Walker has provided a supportive laboratory environment. He is an incredibly kind and patient man. I can't thank him and everyone at the Ragon Institute of MGH, MIT, and Harvard University enough for their help and support in what has been a challenging endeavor. I especially need to thank Doug Kwon, Alicja Piechocka-Trocha, Zaza Ndhlovu, Hendrik Streeck, and Elizabeth Byrne for their assistance, patience, and conversation, many parts of which inspired pieces of this book.

Many researchers helped me with this book—more than I can possibly name here. Their generosity in sharing their personal stories and research, as well as reading and reviewing the manuscript, has been touching. I'd like to particularly thank David Ho, David Baltimore, Anthony Fauci, Carl June, Bruce Levine, John Zaia, Robert Gallo, Steve Deeks, Steve Yukl, Tim Henrich, Dan Kuritzkes, Michael Holmes, and Dave Margolis for taking time out of their busy schedules.

I've been fortunate to have extraordinary teachers in my life. A special thanks goes to Robert Garry, who gave me a foundation in virology and held my hand along the bumpy path to a PhD. I first discovered the joy of DNA replication thanks to my seventh-grade science teacher, David Randall, who sparked my lifelong pursuit of biology. Thanks also go to Michael O'Brien, a teacher who fueled my love of literature.

I wouldn't have been able to write this book without my wonderful family and friends who have given me so much: my parents, Marco Katz and Betsy Boone; my mom, Eva Grundgeiger, and my mother-in-law, Ruby Holt, both of whom I miss dearly, John and Joyce Boone; Ken Holt; Shea Holt; Claire and Jerry McCleery; Sheldon Katz; Rose Grundgeiger; Rachael and Gerry Coakley; Elizabeth Keane; and Sean Cashman. A special thanks to my brother-in-law, Scott Holt, who put his life on hold for weeks at a time to help take care of his niece and without whom this book could not have been written.

The two most important people in my life: my husband, Larkin Holt, who has given me unconditional love, support, and patience when I needed him most, and our daughter, Eleanor Frances Holt, who inspires me to be a better person.

Index

Note: Page numbers in *italics* indicate illustrations.